The publishing house tredition has created the series **TREDITION CLASSICS**. It contains classical literature works from over two thousand years. Most of these titles have been out of print and off the bookstore shelves for decades.

The book series is intended to preserve the cultural legacy and to promote the timeless works of classical literature. As a reader of a **TREDITION CLASSICS** book, the reader supports the mission to save many of the amazing works of world literature from oblivion.

The symbol of **TREDITION CLASSICS** is Johannes Gutenberg (1400 – 1468), the inventor of movable type printing.

With the series, tredition intends to make thousands of international literature classics available in printed format again – worldwide.

All books are available at book retailers worldwide in paperback and in hardcover. For more information please visit: www.tredition.com

tredition was established in 2006 by Sandra Latusseck and Soenke Schulz. Based in Hamburg, Germany, tredition offers publishing solutions to authors and publishing houses, combined with worldwide distribution of printed and digital book content. tredition is uniquely positioned to enable authors and publishing houses to create books on their own terms and without conventional manufacturing risks.

For more information please visit: www.tredition.com

The Boston Terrier and All About It A Practical, Scientific, and Up to Date Guide to the Breeding of the American Dog

Edward Axtell

Imprint

This book is part of the TREDITION CLASSICS series.

Author: Edward Axtell
Cover design: toepferschumann, Berlin (Germany)

Publisher: tradition GmbH, Hamburg (Germany)
ISBN: 978-3-8495-0831-9

www.tredition.com
www.tredition.de

Copyright:
The content of this book is sourced from the public domain.

The intention of the TREDITION CLASSICS series is to make world literature in the public domain available in printed format. Literary enthusiasts and organizations worldwide have scanned and digitally edited the original texts. tradition has subsequently formatted and redesigned the content into a modern reading layout. Therefore, we cannot guarantee the exact reproduction of the original format of a particular historic edition. Please also note that no modifications have been made to the spelling, therefore it may differ from the orthography used today.

Edward Axtell

TABLE OF CONTENTS

- CHAPTER I.
    - The Boston Terrier
- CHAPTER II.
    - The Boston Terrier Club; Its History; The Order of Business; Constitution, By-Laws and Official Standard
    - The Revised Boston Terrier Standard
- CHAPTER III.
    - Kenneling
- CHAPTER IV.
    - General Hints On Breeding
- CHAPTER V.
    - Rearing Of Puppies
- CHAPTER VI.
    - Breeding For Size
- CHAPTER VII.
    - Breeding For Good Disposition
- CHAPTER VIII.
    - Breeding For a Vigorous Constitution
- CHAPTER IX.
    - Breeding For Color and Markings
- CHAPTER X.
    - Sales
- CHAPTER XI.
    - Boston Terrier Type and the Standard
- CHAPTER XII.
    - Picture Taking
- CHAPTER XIII.
    - Notes
- CHAPTER XIV.
    - Conclusion
- CHAPTER XV.
    - Technical Terms Used In Relation To the Boston Terrier, and Their Meaning

INDEX TO ILLUSTRATIONS

- Edward Axtell
- Franz J. Heilborn
- Heilborn's Raffles
- Edward Burnett, a Prominent Early Breeder
- Barnard's Tom
- Hall's Max
- Champion Halloo Prince
- Bixby's Tony Boy
- J. P. Barnard, the Father of the Boston Terrier
- Champion Sonnie Punch
- Rockydale Junior
- Edward Axtell, Jr., and One of His Boston Terriers
- E. S. Pollard, A Large and Successful Breeder
- St. Botolph's Mistress King
- Champion Yankee Doodle Pride
- Champion Dallen's Spider
- Champion Mister Jack
- Champion Caddy Belle
- Prince Lutana
- Champion Fosco
- "Pop" Benson with Bunny II
- Sir Barney Blue
- Champion Lady Dainty
- Champion Todd Boy
- Champion Willowbrook Glory
- Squantum Punch
- Tony Ringmaster
- Goode's Buster
- Champion Whisper
- Champion Druid Vixen
- Champion Remlik Bonnie
- Champion Boylston Reina
- Champion Roxie
- Peter's Little Boy and Ch. Trimont Roman

- Champion Lord Derby
- Gordon Boy, Gretchen, Derby's Buster, Tommy Tucker, Ch. Lord Derby
- Gordon Boy
- Champion Dean's Lady Luana
- Mrs. William Kuback, with Ch. Lady Sensation

CHAPTER I.

THE BOSTON TERRIER.

Return to Table of Contents

Who and what is this little dog that has forced his way by leaps and bounds from Boston town to the uttermost parts of this grand country, from the broad Atlantic to the Golden Gate, and from the Canadian border to the Gulf of Mexico? Nay, not content with this, but has overrun the imaginary borders north and south until he is fast becoming as great a favorite on the other side as here, and who promises in the near future, unless all signs fail, to cross all oceans, and extend his conquests wherever man is found that can appreciate beauty and fidelity in man's best friend. What passports does he present that he should be entitled to the recognition that he has everywhere accorded him? A dog that has in 35 years or less so thoroughly established himself in the affections of the great body of the American people, so that his friends offer no apology whatever in calling him the American dog, must possess peculiar qualities that endear him to all classes and conditions of men, and I firmly believe that when all the fads for which his native city is so well known have died a natural death, he will be in the early bloom of youth. Yea, in the illimitable future, when the historian McCauley's New Zealander is lamenting over the ruins of that marvelous city of London, he will be accompanied by a Boston terrier, who will doubtless be intelligent enough to share his grief. In reply to the query as to who and what he is, it will be readily recalled that on the birth of possibly the greatest poet the world has ever seen it was stated:

"The force of nature could no further go,

To make a third, she joined the other two."

And this applies with equal force to the production of the Boston terrier. The two old standard breeds of world-wide reputation, the English bulldog and the bull terrier, had to be joined to make a third which we believe to be the peer of either, and the superior of both.

The dog thus evolved possesses a type and individuality strictly his own, inherited from both sides of the house, and is a happy medium between these two grand breeds, possessing the best qualities of each. To some the name "terrier" would suggest the formation of the dog on approximate terrier lines, but this is as completely erroneous as to imagine that the dog should approach in like proportion to the bull type. When the dog was in its infancy it was frequently called the Boston bull, and then again the round-headed bull and terrier, and later, when the Boston Terrier Club was taken under the wings of the great A.K.C. in 1893, it became officially known as the Boston terrier.

There are several features that are characteristic of the dog that tend to its universal popularity—its attractive shape, style and size, its winning disposition, and its beautiful color and markings. From the bulldog he inherits a sweet, charming personality, quiet, restful demeanor, and an intense love of his master and home. He does not possess the restless, roving disposition which characterizes so many members of the terrier tribe, nor will he be found quarreling with other dogs. From the bull terrier side he inherits a lively mood, the quality of taking care of himself if attacked by another dog, and of his owner, too, if necessary, the propensity to be a great destroyer of all kinds of vermin if properly trained, and an ideal watch dog at night. No wonder he is popular, he deserves to be. The standard describes him as follows:

"The general appearance of the Boston terrier is that of a smooth, short-coated, compactly built dog of medium station. The head should indicate a high degree of intelligence and should be in proportion to the dog's size; the body rather short and well knit, the limbs strong and finely turned, no feature being so prominent that the dog appears badly proportioned. The dog conveys an impression of determination, strength and activity, style of a high order and carriage easy and graceful."

The men composing the Boston Terrier Club, who framed this standard in 1900, were as thoughtful a body as could possibly be gotten together, and they carefully considered and deliberated over every point at issue, and in my estimation this standard is as near perfect as any can be. I was an interested participant in the discussion of the same, having in my mind's eye as models those two noted dogs owned by that wonderful judge of the breed, Mr. Alex. Goode, Champion Monte, and his illustrious sire, Buster. If one takes the pains to analyze the standard he will be impressed by the perfect co-relation of harmony of all parts of the dog, from the tip of his broad, even muzzle, to the end of his short screw tail. Nothing incongruous in its makeup presents itself, but a graceful, symmetrical style characterizes the dog, and I firmly believe that any change whatever would be a detriment.

Franz J. Heilborn

Heilborn's Raffles

Edward Burnett
A Prominent Early Breeder

It seems to be hardly necessary at this late date to give a history of the dog, but perhaps for that large number of people who are intensely interested in him but have not had the chance to have been made acquainted with his origin, a brief survey may be of service. Although Boston rightly claims the honor of being the birthplace of the Boston terrier, still I think the original start of the dog was in England, for the first dog that was destined to be the ancestor of the modern Boston terrier was a dog named Judge, a cross between an English bull and bull terrier, imported from the other side and owned by Mr. R. C. Hooper, and known as Hooper's Judge.

On my last visit to England I found that quite a number of dogs have been bred in this way, viz., a first cross between the bull and terrier, especially in the neighborhood of Birmingham in the middle of England; but these dogs are no more like the Boston terrier than an ass is like a thoroughbred horse. Judge was a dark brindle, with a white stripe in face, nearly even mouthed, weighing about thirty-two pounds, and approximating more to the bull than the terrier side. He was mated to a white, stocky built, three-quarter tail, low stationed bitch, named Gyp (or Kate), owned by Mr. Edward Burnett of Southboro. Like Judge, she possessed a good, short, blocky head. It may not be out of place to state here that some few years ago, on paying a visit to Mr. Burnett at Deerfoot Farm, Southboro, he told me that in the early days he possessed thirteen white Boston terrier dogs that used to accompany him in his walks about the farm, and woe to any kind of vermin or vagrant curs that showed themselves. From Judge and Gyp descended Well's Eph, a low-stationed, dark brindle dog with even white markings, weighing twenty-eight pounds. Eph was mated to a golden brindle, short-headed, twenty pound bitch, having a three-quarter tail, named Tobin's Kate. From this union came a red brindle dog with a white blaze on one side of his face, white collar, white chest, and white feet, weighing twenty-two pounds, and possessing the first screw tail, named Barnard's Tom. I shall never forget the first visit I made to Barnard's stable to see him. To my mind he possessed a certain type, style and quality such as I had never seen before, but which stamped him as the first real Boston terrier, as the dog is today understood. I was never tired of going to see him and his brother, Atkinson's Toby. Tom was mated to a dark brindle bitch, evenly marked, weighing twenty pounds. She had a good, short, blocky head, and a three-quarter tail, and known as Kelley's Nell. The result of this mating was a dog destined to make Boston terrier history, and to my mind the most famous Boston terrier born, judged by results. He was known as "Mike," commonly called "Barnard's Mike." He was a rather light brindle and white, even mouthed, short tailed dog, weighing about twenty-five pounds, very typical, but what impressed me was his large, full eye, the first I had ever seen, and which we see so often occurring in his descendants. I owned a grandson of his named "Gus," 48136, who was almost a reproduction of him, with eyes fully as large. Unfortunately he jumped out of a third-story window in my kennels and permanently ended his usefulness. Chief among the direct descendants from Hooper's Judge were the noted stud dogs, Ben Butler, Hall's Max, O'Brien's Ross, Hook's Punch, Trimount King, McMullen's Boxer, and Ben, Goode's Ned, and Bixby's Tony Boy. The two dogs that impressed me the most in that group were Max, a fairly good sized, beautiful dispositioned dog that could almost talk, belonging to Dr. Hall, then a house doctor at the Eye and Ear Infirmary, Charles street. He was used, I am told, a great deal in the stud, and sired a great many more puppies than the doctor ever knew of. Bixby's Tony Boy was the other. I had a very handsome bitch by him out of a Torrey's Ned bitch, and liked her so much that I offered Mr. Bixby, I believe, $700 for Tony, only to be told that a colored gentleman (who evidently knew a good thing when he saw it) had offered him $200 more.

Of the line of early bitches of the same breeding may briefly be mentioned Reynold's Famous, dam of Gilbert's Fun; Kelley's Nell, dam of Ross and Trimount King; Saunder's Kate, dam of Ben Butler; Nolan's Mollie, dam of Doctor, Evadne and Nancy.

Quite a number of other small dogs were subsequently introduced into the breed, which had now been somewhat inbred. These were largely imported from the other side, and were similar in type to Hooper's Judge. One of the most noted was the Jack Reede dog. He was an evenly marked, reddish brindle and white, rather rough in coat, three-quarter tail, weighing fourteen pounds. Another very small dog was the Perry dog, imported from Scotland, bluish and white in color, with a three-quarter straight tail, and weighing but six pounds. I have always felt very sorry not to have seen him, as he must have been a curiosity. Still another outside dog, also imported, and very quarrelsome, white in color, weighing eighteen pounds, with a good, large skull, and an eye as full as Barnard's Mike, but straight tail, was Kelley's Brick. Another outside dog (I do not know where he came from), was O'Brien's Ben. He was a short, cobby, white and tan brindle color, three-quarter tail, with a short head and even mouth. It will be observed that practically all these outside dogs were small sized, and were selected largely on that account. By the continued inbreeding of the most typical of the sons and daughters of Tom, the present type of the dog was made permanent.

Barnard's Tom

Hall's Max

Champion Halloo Prince

Bixby's Tony Boy

Perhaps this somewhat restricted review of the breed, going back over thirty-six or seven years and showing the somewhat mixed ancestry of our present blue-blooded Boston terrier of today, may afford some explanation of the diversity of type frequently presented in one litter. I have seen numbers of litters where the utmost attention has been paid to every detail with the expectancy of getting crackerjacks, to find that one will have to wait for the "next time," as the litter in question showed the bull type, and the terrier also, and very little Boston; but fortunately, with the mating intelligently attended to, and the putting aside of all dogs that do not comport to the standard as non-breeders, a type of a dog will be bred true to our highest ideals. My advice to all breeders is, do not get discouraged, try, yes, try again, and Boston terriers, that gladden the eye and fill the pocketbook, will be yours.

CHAPTER II.

THE BOSTON TERRIER CLUB.

ITS HISTORY, THE ORDER OF ITS BUSINESS, CONSTITUTION, BY-LAWS, AND OFFICIAL STANDARD.

Return to Table of Contents

In 1890 a club was formed in Boston by a comparatively small body of men who were very much interested in the dog then known as the Round-Headed Bull and Terrier dog. These men were breeders and lovers of the dog, and their main object in coming together was not to have a social good time (although, happily, this generally took place), but to further the interests of the dog in every legitimate way. The dog had been shown at the New England Kennel Club show, held in Boston in April, 1888, being judged by Mr. J. P. Barnard, Jr., ofttimes styled "the father of the breed," practically two years before the formation of the Club. The year following the Club applied for admission in the American Kennel Club, and recognition for their dogs in the Stud Book. The A. K. C. stated that while perfectly willing to take the Club into its fold, they could not place the dog in the Stud Book, as he was not an established breed, and suggesting, that as the dog was not a bull terrier, and as he was then bred exclusively in Boston, the name of the "Boston Terrier Club." The year following the A. K. C., after a great deal of persuasion by the loyal and devoted members of the Club, became convinced of the merits of the breed, and formally acknowledged the same by admitting the Club to membership, and giving their dog a place in the official Stud Book.

The Boston Terrier Club is duly incorporated under the laws of Massachusetts, has a present membership of from seventy-five to a hundred, men and women who are devoted to the dog, and willing to do everything for its advancement. The annual meeting is held on the second Wednesday in December, at which a number of judges are elected, whose names are forwarded to the bench show committees of the principal shows, requesting that one of the number be elected to officiate as judge of the Boston terriers. Monthly meetings are held which are always exceedingly interesting and instructive.

The officers are elected by printed ballots sent to all members of the Club, who mark and return them. They consist of the president, vice-president, secretary, and treasurer. The executive committee consists of the officers (ex officio) and three others.

The Club gives a specialty show yearly in Boston and is the largest and greatest of one breed fixtures; the dog being, in fact, one of the largest supporters of the dog shows in the country. Cups and medals are offered at most of the bench shows for competition among the members, and at the Ladies' Kennel Association shows a cup and medal were offered, open to all exhibitors of Boston terriers.

In view of the fact that so many Boston Terrier Clubs are starting up all over the country, and even beyond, the following Order of Business, Constitution, By-Laws, and Official Standard, can safely be taken as models:

ORDER OF BUSINESS.

1. Calling meeting to order.
2. Roll call.
3. Reading of minutes.
4. Reports of officers.
5. Reports of standing committees by seniority.
6. Reports of special committees.
7. Communications.
8. Applications for membership.
9. Election of members.
10. Election of officers.
11. Unfinished business.
12. New business.
13. Welfare of the Club.
    - Under this heading is included remarks and debates intended to promote the interests of the Club and the Boston terrier in general.
14. Adjournment.

CONSTITUTION.

ARTICLE I.

NAME.

This Association shall be known as and called the Boston Terrier Club.

ARTICLE II.

OBJECT.

The object of the Club shall be to promote and encourage the breeding and improvement of the Boston Terrier Dog, as defined by its standard.

ARTICLE III.

MEMBERSHIP.

Section 1. Applications for membership must be accompanied by the membership fee and endorsed by two members, and made at least seven days before action by the Club, to the secretary or a member of the membership committee, who shall refer it to said committee for investigation.

Sec. 2. Any member can resign from the Club by sending his resignation to the secretary in writing, and upon the acceptance of such, all his interest in the property of the Club ceases from the date of such resignation.

Sec. 3. Any member whose dues shall remain unpaid for one month after the same becomes due, shall cease to be a member, and forfeit to the Club all claims and benefits to which he would have been entitled as a member, provided that the executive committee may consider his case, and upon sufficient cause shown, reinstate him to membership upon payment of his dues.

ARTICLE IV.

MANAGEMENT.

Section 1. The officers of the Club shall consist of a president, vice-president, secretary, treasurer, and an executive committee, of which three shall constitute a quorum; said committee to consist of the above named officers and three active members chosen by the Club.

Sec. 2. Any office vacated during the year shall be filled by the executive committee.

ARTICLE V.

Section 1. Nomination for officers and judges for the ensuing year shall be made either by mail or from the floor, at a meeting to be held in November, at least twenty days prior to the annual meeting, the call to contain the purpose of the meeting, after which nominations shall be closed. The secretary shall mail a ballot containing all regular nominations to each member in time to be voted at the annual meeting.

Sec. 2. The officers of the Club shall be chosen by ballot at the annual meeting and shall hold their respective offices for one year or until their respective successors are elected.

Sec. 3. Mail voting shall be allowed on amendments to the Constitution, By-Laws, Standard and Scale of Points.

Sec. 4. Each member shall have the right to vote on the election of officers and judges by mailing the official ballot duly marked and sealed to the secretary, and enclosed in an envelope, which envelope shall also contain the name of the member so voting.

ARTICLE VI.

MEETINGS.

Section 1. There shall be meetings of the Club, at which seven members present and voting shall constitute a quorum, held at Boston, Mass., at such time and place as the president may direct, but the annual meeting shall be held on the second Wednesday in December of each year.

SPECIAL MEETINGS.

Sec. 2. A special meeting of the Club shall be called by the president on the written application of five members in good standing.

BY-LAWS.

ARTICLE I.

DUTIES OF OFFICERS.

Section 1. President. — The president shall discharge the usual duties of his office, preside at all meetings of the Club and of the executive committee, call special meetings of the Club, or of the executive committee, and enforce the provisions of the Constitution and By-Laws of the Club. He may vote on amendments to the Constitution or alteration of the By-Laws and Standard or Scale of Points, on the expulsion or suspension of a member, and on election of officers and judges. But on all other matters he shall vote only in case of tie and then give the deciding vote.

Sec. 2. Vice-President. — The vice-president shall discharge all the duties of the president in the latter's absence.

Sec. 3. Secretary. — The secretary shall have charge of all official correspondence, keep copies of all letters sent by him, and file such as he may receive, and correspond at the request of the president or executive committee on all matters appertaining to the object of the Club. He shall keep a roll of the members of the Club with their addresses.

He shall be exempt from payment of annual dues.

Sec. 4. Treasurer. — The treasurer shall collect and receive all moneys due the Club and keep a correct account of the same. He shall pay all orders drawn on him by the executive committee out of the funds of the Club, when countersigned by the president, and present a report of the condition of affairs in his department at the request of the executive committee or president, and at the annual meeting. The treasurer shall furnish a bond satisfactory to the executive committee.

Sec. 5. Committees. — The executive committee shall make all purchases ordered by the Club, audit the accounts of the treasurer and report the same at the annual election in December, and transact all business not otherwise provided for.

It shall have the power to appoint sub-committees for any special purpose, and to delegate to each sub-committee the powers and functions of the committee relating thereto.

The president shall be the chairman of the executive committee.

Sec. 6. Sub-Committees. — The standing sub-committees shall be a membership committee of five and a pedigree committee of three.

The membership committee shall investigate the standing of all applicants, and report to the Club for action those names it considers as desirable members.

The pedigree committee shall investigate the pedigrees of those dogs offered for registration in the Boston Terrier Stud Book.

The chairman of the pedigree committee shall have the custody of the Club stud book, and shall enter in the same the registrations allowed by the B. T. C.

ARTICLE II.

DISCIPLINE.

The executive committee shall have the power to discipline by suspension a member found guilty of conduct prejudicial to the best interests of the Club. All charges against a member must be made in writing and filed with the executive committee, and no member shall be suspended without an opportunity to be heard in his own defense. When the expulsion of a member is considered advisable, the report of the committee shall be presented to the Club, whose action shall be final.

ARTICLE III.

DUES.

Section 1. The entrance fee shall be five dollars, which must accompany the application for membership.

Sec. 2. The annual dues shall be ten dollars, payable upon notice of election and at each annual meeting thereafter.

ARTICLE IV.

JUDGES.

Section 1. There shall be elected by ballot each year at the annual meeting a corps of not more than fifteen judges, a list of whose names shall be sent to bench show committees with a request that the judge of Boston terriers at their approaching shows be selected from said list.

Sec. 2. The Club judges may exhibit, but shall not compete at or be interested directly or indirectly in the show at which they officiate.

ARTICLE V.

AMENDMENTS.

This Constitution and these By-Laws, and the Standard and Scale of Points may be amended or altered by a two-thirds vote at any regular meeting or special meeting called for that purpose.

Notice of proposed change having been given to all members at least ten days previous to said meeting.

THE REVISED BOSTON TERRIER STANDARD

The present Boston terrier standard was adopted by the Boston Terrier Club on October 7, 1914, as a result of a revision recommended by a committee appointed by the Boston Terrier Club.

It was felt, in view of the fact that the dog had become established all over the continent among breeders and fanciers not as familiar with the ideal of the breed as were the original breeders and friends of the dog around Boston, that a more explicit, definite standard, one that could be more easily understood by the great body of the dog's admirers of today, should be adopted.

It will be readily observed by a comparison of the old standard, which has practically been in existence since the formation of the club in 1891, that no vital point has been really changed.

REVISED STANDARD		OLD STANDARD.
Point Values		Scale of Points.
	GENERAL APPEARANCE: The general appearance of the Boston terrier should be that of a lively, highly intelligent, smooth coated, short headed, compactly built, short tailed, well balanced dog of medium station, of brindle color and evenly marked with white. The head should indicate a high degree of intelligence and should be in proportion to the size of the dog; the body rather short and well knit, the limbs strong and neatly turned; tail short and no feature being so prominent that the dog appears badly proportioned. The dog should convey an impression of determination, strength	GENERAL APPEARANCE AND STYLE: The general appearance of the Boston Terrier is that of a smooth, short-coated, compactly-built dog of medium station. The head should indicate a high degree of intelligence and should be in proportion to the dog's size; the body rather short and well-knit, the limbs strong and finely turned, no feature being so prominent that the dog appears badly proportioned. The dog conveys an impression of determination, strength and activity. Style of a high order, and carriage easy
10		10

and activity, with style of a high order; carriage easy and graceful. A proportionate combination of "Color" and "Ideal Markings" is a particularly distinctive feature of a representative specimen, and dogs with a preponderance of white on body, or without the proper proportion of brindle and white on head, should possess sufficient merit otherwise to counteract their deficiencies in these respects.

The ideal "Boston Terrier Expression" as indicating "a high degree of intelligence," is also an important characteristic of the breed.

"Color and Markings" and "Expression" should be given particular consideration in determining the relative value of "General Appearance" to other points.

and graceful.

SKULL: Square, flat on top, free from wrinkles; cheeks flat; brow abrupt, stop well defined.
1
2

SKULL: Broad and flat, without prominent cheeks, and forehead free from wrinkles.
1
2

STOP: Well defined, but indenture not too deep.
2

EYES: Wide apart, large and round, dark in color, expression alert, but kind and intelli-
5

EYES: Wide apart, large and round, neither sunken nor too prominent, and in
5

gent; the eyes should set square across brow and the outside corners should be on a line with the cheeks as viewed from the front.

MUZZLE: Short, square, wide and deep; free from wrinkles; shorter in length than in width and depth, and in proportion to skull; width and depth carried out well to end. Nose black and wide, with well defined line between nostrils. The jaws broad and square, with short regular teeth. The chops of good depth, but not pendulous, completely covering the teeth when mouth is closed. The muzzle should not exceed in approximate length one-third of length of skull.

¹
²

EARS: Small and thin, situated as near corners of skull as possible.

HEAD FAULTS: Skull "domed" or inclined; furrowed by a medial line; skull too long 2 for breadth, or vice versa; stop too shallow; brow and skull too slanting. Eyes small or sunken; too prominent; light color; showing too much white or haw. Muzzle wedge shaped or lacking depth; down faced; too much cut out below the

color dark and soft. The outside corner should be on a line with the cheeks as viewed from the front.

MUZZLE: Short, square, wide and deep, without wrinkles. Nose black and wide, with a well defined straight line between nostrils. The jaws broad and square, with short, regular teeth. The chops wide and deep, not pendulous, completely covering the teeth when mouth is closed.

¹
²

EARS: Small and thin, 2 situated as near corners of skull as possible.

eyes; pinched nostrils; protruding teeth; weak lower jaw; showing "turn up." Poorly carried ears or out of proportion.

NECK: Of fair length, slightly arched and carrying the head gracefully; setting neatly into shoulders. 3

NECK FAULTS: Ewe-necked; throatiness; short and thick.

NECK: Of fair length, without throatiness and slightly arched. 5

BODY: Deep with good width of chest; shoulders sloping; back short; ribs deep and well sprung, carried well back of loins; loins short and muscular; rump curving slightly to set-on of tail. Flank slightly cut up. The body should appear short, but not chunky. 1 / 5

BODY FAULTS: Flat sides; narrow chest; long or slack loins; roach back; sway back; too much cut up in flank.

BODY: Deep and broad of chest, well ribbed up. Back short, not roached. Loins and quarters strong. 1 / 5

ELBOWS: Standing neither in nor out. 4

ELBOWS: Standing neither in nor out. 2

FORELEGS: Set moderately wide apart and on a line with the points of the shoulders; straight in bone and well muscled; pasterns short and strong. 5

FORELEGS: Wide apart, straight and well muscled. 4

34

HINDLEGS: Set true; bent at stifles; short from hocks to feet; hocks turning neither in nor out; thighs strong and well muscled. — 5

FEET: Round, small and compact, and turned neither in nor out; toes well arched.

LEG AND FEET FAULTS: Loose shoulders or elbows; hind legs too straight at stifles; hocks too prominent; long or weak pasterns; splay feet. — 5

TAIL: Set-on low; short, fine and tapering; straight or screw; devoid of fringe or coarse hair, and not carried above horizontal.

TAIL FAULTS: A long or gaily carried tail; extremely gnarled or curled against body. — 5

(Note: The preferred tail should not exceed in length approximately half the distance from set-on to hock.)

COLOR: Brindle with white markings. — 4

HINDLEGS: Straight, quite long from stifle to hock (which should turn neither in nor out), short and straight from hock to pasterns. Thighs well muscled. Hocks not too prominent. — 4

FEET: Small, nearly round, and turned neither in nor out. Toes compact and arched. — 2

TAIL: Set-on low, short, fine and tapering, devoid of fringe or coarse hair, and not carried above the horizontal. — 10

COLOR: Any color, brindle, evenly marked with white, strongly preferred. — 8

IDEAL MARKINGS: White muzzle, even white blaze over head, collar, breast, part or whole of forelegs and hindlegs below hocks.

10 COLOR AND MARKINGS FAULTS: All white; absence of white markings; preponderance of white on body; without the proper proportion of brindle and white on head; or any variations detracting from the general appearance.

4 MARKINGS: White muzzle, blaze on face, collar, chest and feet.

3 COAT: Short, smooth, bright and fine in texture.
COAT FAULTS: Long or coarse; lacking lustre.

3 COAT: Fine in texture, short, bright and not too hard.

100

100

WEIGHTS: Not exceeding 27 pounds, divided as follows:

- Lightweight: Under 17 pounds.
- Middleweight: 17 and not exceeding 22 pounds.
- Heavyweight: 22 and not exceeding 27 pounds.

WEIGHT: Lightweight class, 12 and not to exceed 17 pounds; middleweight class, 17 and not to exceed 22 pounds; heavyweight class, 22 and not to exceed 28 pounds.

DISQUALIFICATIONS: Solid black, black and tan, liver

DISQUALIFICATIONS: Docked tail and any artifi-

and mouse colors. Docked tail and any artificial means used to deceive the judge.

cial means used to deceive the judge.

J. P. Barnard
The Father of the Boston Terrier

Champion Sonnie Punch

Rockydale Junior

AN EARLY STANDARD

The following standard adopted when the dog was known as the Round-Headed Bull and Terrier Dog, will be of interest here.

- Skull—Large, broad and flat.
- Stop—Well defined.
- Ears—Preferably cut, if left on should be small and thin, situated as near corners of skull as possible; rose ears preferable.
- Eyes—Wide apart, large, round, dark and soft and not "goggle" eyed.
- Muzzle—Short, round and deep, without wrinkles, nose should be black and wide.
- Mouth—Preferably even, teeth should be covered when mouth is closed.
- Neck—Thick, clean and strong.
- Body—Deep at chest and well ribbed up, making a short backed, cobby built dog; loins and buttocks strong.
- Legs—Straight and well muscled.
- Feet—Strong, small and moderately round.
- Tail—Short and fine, straight or screw, carried low.
- Color—Any color, except black, mouse or liver; brindle and white, brindle or whole white are the colors most preferred.
- Coat—Short, fine, bright and hard.
- Symmetry—Of a high order.
- Disqualifications—Hair lip, docked tail and any artificial means used to deceive the judge.
- Weight—It was voted to divide the different weights into three classes, as follows: 15 pounds and under, 25 pounds and under, 36 pounds and under.

Scale of points:

Skull	15
Muzzle	15

Nose	5
Eyes	5
Ears	5
Neck	5
Body	10
Legs and Feet	10
Tail	10
Color and Coat	10
Symmetry	10
Total	100

CHAPTER III.

KENNELING

Return to Table of Contents

It goes without saying that any place is not good enough for a dog, although when one considers the way some dogs are housed in small, dark outbuildings, or damp, ill-lighted and poorly ventilated cellars, or even perhaps worse, in old barrels or discarded drygoods boxes in some out-of-the-way corner, it is not surprising the quality of the puppies raised in them.

A great many people who only keep one or two dogs keep them in the kitchen or living room, and here, of course, conditions are all right, but the fancier who keeps any considerable number will find that it pays to house his dogs in a comfortable, roomy, dry building, free from draughts, on high lands (with a gravel foundation, if possible), that can be flooded with sunshine and fresh air. Such a kennel can be simple or elaborate in construction, severely plain or ornamental in its architecture, but it must possess the above characteristics in order to have its occupants kept in the pink of condition. Where half a dozen dogs are kept, I think a kennel about 20 feet long, nine feet wide, with a pitched roof, nine feet high in the front, and at the back seven feet, with a southern exposure, with good windows that open top and bottom, and a good tight board floor will do admirably. This can, of course, be partitioned off in pens to suit, with convenient runs outside wired at the top to prevent dogs jumping over. The building should, of course, be well constructed, covered with good sheathing paper, and either clapboarded or shingled. Such a building should be cool in summer and warm in winter, and thoroughly weather proof. If provided with a good "Eureka ventilator" and well painted, the dogs and their owner will be satisfied. Where a much larger number of dogs are kept, then a corresponding amount of floor space is a necessity. I rather like the style of a kennel, say from fifty to a hundred feet long, twelve to fifteen feet wide, with an open compartment or shed, about twelve feet long (in which the dogs can take a sun bath or get the air if the weather is not favorable to go outside. This also makes an ideal feeding pen), in the middle of the house, without outside runs to each pen, and each run opening into a large exercising yard, so that all the dogs may have a good frolic together, of course, under the watchful eye of the kennel man.

The large breeders will also require a separate building at some distance from the main kennels for use as a hospital, a small kennel for his bitches in season, and some small, portable kennels which can be placed under adequate shade trees for his litters of puppies during the hot weather. It would be an excellent plan if good shade trees could be planted to cover all the runs, but if this is not possible, then it is advisable to have at the rear of the kennels a clear space covered over with a roof, say ten or twelve feet wide, for the dogs to have free access to during the heat of the day.

Perhaps a description of our own kennels, entirely different in construction from these, and costing more to build, may be of interest here. We have two buildings, seventy-five feet apart, built exactly like a house, with two stories and a high basement or cellar, twenty-five feet wide and thirty feet long. One of these houses is lined with matched paneling and divided off on each floor into separate compartments; the other is only boarded, one thickness of good paper and clapboarded and, of course, not nearly as warm. This second building has no pens in it. The basement has a stone wall at the back, but on the east, south and west sides is boarded to the ground, and has a dry gravel floor. These buildings are well supplied with windows (the same as a house), and get the sun all day. In these buildings we have no artificial heat whatever, and all stock, except small puppies, are kept there. Our pups in the winter have warm quarters until they are four months old, when they are placed in the south side of the warmer kennels. All puppies are kept in the cool basement in the hot weather, and during the summer our bitches in whelp are kept there also. We have not any separate runs attached to these buildings, which entails a much closer watch on the dogs, of course, but each building opens into a very large enclosure with abundant shade trees, and the dogs can, if let out, have the run of several acres.

In the fall of the year we have several tons of rowen (second crop hay with a good deal of clover in it) put in the upper story of the open kennel, and a smaller amount in the first story, and during the winter a certain number of young dogs that will not quarrel amongst themselves are given the run of the building where they burrow into the soft hay and are as comfortable as can be. Particular care has to be taken that they do not get any bones or any food to quarrel over, or trouble would ensue right away. Allow me to say that only dogs brought up together with perfect dispositions can be allowed to run together. A strange dog must never be placed with them or his days will be numbered. In the summer, of course, no dogs are kept in the upper story, as they would suffer from the heat. Also no bitches in whelp are ever allowed to run together.

In the other kennel in each pen during the cold weather is a large, tight box, with hole in side, filled with this soft hay, renewed when necessary, in which two dogs sleep very comfortably. The windows in each kennel, as soon as the weather permits, are kept open at the top night and day, and top and bottom while the dogs are out doors in the daytime, and in this way the kennels can be kept perfectly sweet and sanitary. Three times during the year, in spring, mid-summer and fall, the kennels are treated with a thorough fumigation of sulphur. We buy bar sulphur by the barrel of a wholesale druggist or importer, and use a good quantity (a small dose does not do much good), keeping the kennel windows and doors tightly closed for twelve hours, after which the building is thoroughly aired before the dogs are returned. Of course, this would not be practical during the winter, nor is it at all necessary. We find that once a week (except of course, during the cold weather), it is a good plan to give the woodwork that the dog comes in contact with a good sprinkling with a watering pot with a solution of permanganate of potassium, using a tablespoonful of the crystals dissolved in a quart of hot water. It costs at wholesale fifty cents per pound, and is the best disinfectant I have ever used. Unless the kennels are kept scrupulously clean the dogs' eyes, especially the puppies, are liable to become seriously inflamed. The gravel in the basement we remove to a depth of eight inches twice a year, putting fresh in its place. Where a large number of dogs are kept it will be found very convenient to have a cook house, wash room and a small closet for kennel utensils in close proximity to the kennels.

By attending to these important essentials, viz., an abundance of pure air and sunshine, protection from dampness, draughts, and cold, proper disinfecting, and sufficient protection from the intense heat of summer, good health, and a reasonable amount of success can be confidently expected, but disease will surely find an entrance where these requirements are not met.

I would like to add that kennels only large enough for white mice, or perchance piebald rats, can never be successfully used to raise Boston terriers in.

CHAPTER IV.

GENERAL HINTS ON BREEDING.

Return to Table of Contents

Having become possessed of suitable kennels to house his stock, the breeder is confronted with the great question: How and where shall I obtain my breeding stock? Much depends on a right start and the getting of the proper kind of dogs for the foundation. Our celebrated Boston poet, Oliver Wendell Holmes, when asked when a child's education should begin, promptly replied, "A hundred years before it was born." This contains an inherent truth that all breeders of choice stock of whatever description it may be, recognize. To be well born is half the battle, and I think this applies with particular force to the Boston terrier, for without a good ancestry of well bred dogs, possessing the best of dispositions, constitutions and conformity to the standard, he is worse than useless.

Whether the start is made with one bitch or a dozen, I believe the best plan to follow is to obtain of a reliable breeder, noted for the general excellence of his dogs in all desirable characteristics, what he considers the best stock obtainable for breeding purposes. This does not imply, of course, that these bitches will be candidates for bench honors, but it does mean that if mated with suitable sires the production of good, all-round puppies with a reasonable amount of luck will be the result. It would be useless to attempt to deal with the subject of breeding in more than a few of its aspects, for after a period of twenty-five years of expended and scientific experiments in the breeding exclusively of Bostons, I shall have to confess that there are many problems still unsolved. The rules and regulations that govern the production of many other breeds of dogs seem impotent here, the assumption that "like produces like" does not seem to hold good frequently in this breed, but perhaps the elements of uncertainty give an unspeakable charm to the efforts put forth for the production of the dogs which will be a credit to the owner's kennel. The old adage that "there is nothing duller than a *puzzle* of which the answer is known," can readily be applied here. I shall endeavor to confine my remarks to the laws observed and the lines followed for the production of dogs in our kennels, especially in the attainment of correct color and markings, vigorous constitutions and desirable dispositions.

In speaking of the breeding stock I am aware that I am going contrary to the opinion of many breeders when I state that I believe that the dam should possess equal or more quality than the sire, that her influence and characteristics are perpetuated in her posterity to a greater degree than are those of the sire's, especially that feature of paramount importance, a beautiful disposition, hence I speak of the maternal side of the house first. There are two inexorable laws that confront the breeder at the onset, more rigid than were those of the Medes and Persians, the non-observance of which will inevitably lead to shipwreck. Better by far turn one's energies in attempting to square the circle, or produce a strain of frogs covered with feathers, than attempt to raise Boston terriers without due attention being given to those physiological laws which experience has proven correct. The first law is that "Like produces like," although, as previously stated in the case of this breed, more than in any other known to the writer, many exceptions present themselves, even when the utmost care has been exercised, still the maxim holds good in the main. The second law is that of Heredity, too often paid inadequate attention to, but which demands constant and unremitting apprehension, as it modifies the first law in many ways. It may be briefly described as the biological law by which the general characteristics of living creatures are repeated in their descendants. Practically every one has noticed its workings in the human family, how many children bear a stronger resemblance to their grandparents, uncles, cousins, etc., than to their parents, and in the lower order of animals, and it seems to me in the Bostons especially, this tendency to atavism, or throwing back to some ancestor, in many cases quite remote, is very pronounced, hence the necessity of a good general knowledge of the pedigree and family history of the dogs the breeder selects for his foundation stock. A kennel cannot be built in a day; it takes time, money, perseverance, and a strict attention to detail to insure success.

"Breed to the best," is a golden rule, but this applies not only to the animals themselves, but also in a far greater measure to the good general qualities possessed by their ancestry. Far more pregnant with good results would be the mating of two good all-round specimens, lacking to a considerable extent show points, but the products of two families known for their general excellence for several generations, than the offspring would be of two noted prize winners of uncertain ancestry, neither of which possessed the inherent quality of being able to reproduce themselves. It will be noted that very few first prize winners had prize winning sires and dams. The noted stud dogs of the past, "Buster," "Sullivan's Punch," "Cracksman," "Hickey's Teddy IV." and many others were not in themselves noted winners, and the same statement may be made of the dams of many of the prize winning dogs, but they possessed in themselves and their ancestry that "hall mark" of quality which appeared in a pronounced form in their offspring. Experience has shown that first class qualities must exist for several generations in order to render their perpetuation highly probable. The converse of this is equally true, that any bad qualities bred for the same length of time are quite as hard to eliminate. If the dog or bitch possesses weak points, be sure to breed to dogs coming from families that are noted for their corresponding strong points. In this case the principle of "give and take" will be adopted. It used to be the ambition of every breeder (or, at least, most of them), to produce a winner, rather than the production of a line of dogs of good uniform type, of good average salable quality, but most have lived long enough to see that this has not paid as well in money or expected results as where similar endeavors have been directed towards the production of good all-round dogs, always striving to advance their dogs to a higher grade of excellence. In this way in nearly every instance prize winning dogs have been produced, and there is this peculiarity noticeable in this breed, that any one, whether he be a breeder of the greatest number, or a very poor man owning only one or two in his kitchen kennel, possesses an equal chance of producing the winner of the blue. The breeder of today has a far easier time than in the early days of the dog when type was not as pronounced or fixed, and when considerable inbreeding of necessity had to be resorted to. In almost all parts of the country stud dogs of first class lineage are obtainable and the general public are educated sufficiently to understand the good points of the dog. I think the breeding of this dog appeals to a wider class of people than any other breed, from the man of wealth who produces the puppies to be given away as wedding presents or Christmas gifts, down to the lone widow, or the man incapacitated for hard work, who must do something to keep the wolf from the door, and who finds in the raising of these charming little pets a certain source of income and a delightful occupation combined. I do not think that any one may apprehend that the market will ever be overstocked, for as the dog becomes known, the desire for possession among all classes will be correspondingly increased, and as he is strictly an American product, no importation from Europe can possibly supply winners, or specially good dogs, as is the case with almost all other breeds. And the fact is demonstrated that dogs of A 1 quality can be produced on American soil.

There are two or three subjects that demand the most careful consideration at the hands of the breeder, and to which I am afraid in many cases not particular enough attention is given. I refer in the first place to the question of inbreeding, an admitted necessity in the early history of the dog, but in the writer's estimation very harmful and much to be discouraged at the present time. I will yield to no man in the belief that the fact is absolutely and scientifically true that close consanguineous breeding is the most powerful means of determining character and establishing type, in many instances justifiable as the only correct way to fix desirable qualities, both physical and mental, but extreme care must be exercised that both parties to the union must be of good quality and not share the same defects, and where it is evident that the extra good qualities on the one side more than outbalance the defects of the other, and extreme precaution must always be paid to avoid carrying this system too far.

In regard to intense inbreeding, as in the case of mating dogs from the same sire and dam, or the bitch to her sire, or dam to son, I thing it is highly objectionable and should never under any circumstances be resorted to; failure will ensue. Far better to let the bitch go by unmated and lose six months than mate her in this way because a suitable stud dog was not at the time available. I believe that this inbreeding is productive of excessive nervousness, weakness in physical form, the impairment of breeding functions, and the predisposition to disease in its multiform manifestations.

Edward Axtell, Jr.,
and One of His Boston Terriers

E. S. Pollard,
A Large and Successful Breeder

St. Botolph's Mistress King

That eminent authority, Sir John Seabright, the originator of the early race of bantams, known as the silver and gold spangled Seabrights, also conducted an exhaustive series of experiments on the inbreeding of dogs and demonstrated to an absolute certainty that the system was productive of weakness, diminished growth, and general weediness. His experiments had a world-wide reputation and the writer, when he first visited his large estates near London, little dreamed that in after years he would personally benefit by Sir John's work. I believe the prevailing ideas in many quarters a number of years ago, as to the general stupidity of the Boston terrier (and in some isolated cases I believed well founded), arose from the fact that it was popularly believed he was too much inbred. I will give just one case of inbreeding in our kennels, tried for experiment's sake, as a warning. I took the most rugged bitch I possessed and mated her to her sire, a dog of equal vigor. The result was six puppies, strong, and as handsome as a picture. When two months old they were sold to different parties on the Eastern seaboard, from Philadelphia up to the Canadian line. This was before the West had "caught on" to the breed. About two months later I had a letter from New York stating that the pup was growing finely, but that he seemed to be hard of hearing. A few days after this I received another epistle from Salem that the puppy I had sent on was believed to be stone deaf. It would be superfluous to add that the purchase money was returned, and the other four customers were notified of the condition of the others. It may seem somewhat incredible, but two out of the four stated that they believed the pups had defective hearing, and declined to receive their money back, and the other two stated that before my notification they had never observed that their dogs were deaf. Here was a case of the entire litter being perfect practically in every other respect, and yet every one stone deaf, and in my estimation not worth a sou. As we have never had a case of deafness in our kennels before or since, we attribute this solely to inbreeding.

Another important feature, little understood, and frequently much dreaded, is that of Antecedent Impressions. When a bitch has been served by a dog not of her own breed it has been proven in extremely rare cases that the subsequent litters by dogs of her own kind, showed traces (or, at least, one or more of the litter did) of the dog she was first lined by. The theory by physiologists is that the life-giving germ, implanted by the first dog, penetrates the serous coat of the ovary, burrows into its parenchyma, and seeks out immature ova, not to be ripened and discharged perhaps for years, and to produce the modifying influence described. Many breeders are unwise enough to believe that a bitch the victim of misalliance is practically ruined for breeding purposes and discard her. While, of course, we believe in the fact of Antecedent Impressions, we think they are as rare as the proverbial visit of angels. We have given this subject serious attention and have tried numerous experiments, using various dogs to ward our bitches, including a pug, spaniel, wire-haired fox terrier, pointer, and perhaps one other, and we have never seen a trace of these matings in subsequent litters. One case, for example: In another part of this book we allude to a dog spoken of by Dr. Mott, in his "Treatise of the Boston Terrier," named "Muggy Dee." The grandmother of this charming little dog was bred in our kennels, by name, "St. Botolph's Bessie." We sold her to a Boston banker, and she matured into a beautiful dog. Upon coming in season she was unfortunately warded by a spaniel on the estate, which so disgusted her owner that he gave her to the coachman. She proved a perfect gold mine to him, as she raised two litters of elegant ideal Bostons every twelve months for a great number of years, and never at any time showed any result of the misalliance.

On the subject of Mental Impressions we need say but little, as the chances of it ever taking place are so small that we merely give it a passing notice and say that in all our experience we have never been troubled with a case. For the benefit of the uninitiated will briefly state that this consists of the mental impression made on the mind of a bitch by a dog with whom she has been denied sexual intercourse, affecting the progeny resulting from the union of another dog with the bitch, generally in regard to the color, and this strange phenomena, when it does occur, is apt to mark usually one puppy of each litter.

A fact not generally known by breeders is that if a bitch is lined by a second dog at any time during heat, the chances are that a second conception may take place, resulting in two distinct sets of pups, half-sister or brother to each other. This fact we have proven.

There is one other important feature which must be noticed before this chapter is closed, and that is Predetermining the Sex. Most breeders, of course, are anxious to have male pups predominate in a litter, and it is a demonstrated fact that ordinary mating produces from four to ten per cent more males than females. For a number of years I had always believed it was impossible to breed so as to attain more than the excess of males above noted, but several years ago I accepted an invitation from Mr. Burnett, of Deerfoot Farm, of Southboro (the owner of Kate or Gyp, the mother of the breed), to spend the day. He was, as will be recalled, one of the earliest and most enthusiastic breeders of the Boston, and is now a scientific breeder of choice dairy stock. We had been discussing a number of problems in regard to raising stock, when he exclaimed: "Mr. Axtell, I believe I have discovered the problem of sex breeding. If I want heifer calves, I breed the cow as soon as she comes in season. If a bull calf is wanted, the cow is served just before going out of season." And said he, "In nineteen experiments I have only been unsuccessful once, and I think you might try the same plan with your Bostons." I have since done so, and although not nearly the same measure of success has attended my experiments as his, yet by breeding bitches at the close of the heat rather than at its commencement, the number of males in a litter has materially increased. Again, I find if a young, vigorous dog is bred to a similar bitch, females will predominate in the offspring, whereas, if the same bitch is bred to a much older dog, an excess of males will generally occur. Occasionally some dogs will be met with that no matter what mated with, will produce largely males, and some the opposite of this will nearly always produce females, and some bitches, no matter how bred, do likewise, but these are exceptions, and not the rule. A kennel man need never worry about sex, inasmuch as good dogs of either gender will always be in demand.

The law of Selection must be carefully attended to to insure the best results. Choose your best and most typical bitches for breeding, especially those that approximate rather to the bull type and are rather long in body and not too narrow in their hind quarters. I do not care if the dam has a somewhat longer tail than the dog, my experience has been that a bitch possessing a tight screw tail did not do quite as well in whelping as one having one a little longer. Do not consider this as suggesting that the tail is a matter of secondary importance, by no means, it is of primal import, and too much attention can never be given to the production of this distinguishing mark of the dog. A Boston without a good tail is almost as worthless as a check without a signature.

Be sure at the time of breeding the bitch is free from worms. A great many are troubled whose owners are totally ignorant of the fact, and this frequently accounts for non-success. Always remember that worms thrive the most when the alimentary canal is kept loaded with indigestible or half-digested food, and that liquid foods are favorable to these pests, while solids tend to expel them. Freshly powdered areca nut, in teaspoonful doses, and the same quantity of a mixture of oil of male fern and olive oil, three parts oil and one part male fern oil, I find are both excellent vermifuges to give to matured dogs. Give a dose and two days after repeat, and this, I think, will be found generally effectual.

Do not, on any account, allow the breeding stock to become too fat. Proper feeding and exercise, of course, will prevent this. It will be found if this is not attended to that the organs of generation have lost their functional activity, and if pups are produced, are, as a rule, small and lack vigor. My experience with Bostons is that it is very desirable to breed them as often as they come in season; if allowed to go by it will be found increasingly harder to get them in whelp. I think a stud dog, to last for a reasonable number of years, should not be used more frequently than once a week. I have found it pays best to give the bitch in whelp a generous feed of raw meat daily. It often effectually prevents the puppy-eating habit.

In closing these general hints on breeding, allow me to say there is no reason whatever, if one has a genuine love for the dog and is thoroughly in earnest in his attentions to it, why the breeding problem should possess any great terrors for him. Perhaps, before closing this chapter, it might be well to write on one or two matters, practically of no special import, but which may at times be instructive and illuminate some few incidents that may puzzle the beginner.

I allude first to that strange phenomena known as "false heat," to which Bostons, more than any other breed with which the writer is familiar, are liable, and which consists of the bitch coming "in season" between the two periods in the year when she legitimately should do so, and after being warded by the dog, is, of course, not in whelp. The next is somewhat akin to this, and consists of the fact that the bitch, after being properly warded by a dog, notwithstanding all the external evidences of being in whelp, even to the possession of milk in her breasts at the expiration of the ninth week, is not so, neither has she been. If, in addition to the above symptoms, and there has been unusual abdominal, uterine, and breast enlargement, with a discharge of blood for several days and no pups are in evidence, then in this case it may safely be concluded that the offspring fell victims to the puppy-eating habit, in which case a close watch must be kept on the bitch at the next time of whelping, as this is a curable habit generally. I have had two cases to my knowledge, both of which were cured I think, largely by giving these two bitches all the raw meat they could possibly eat while in whelp. One other fact, related somewhat to the last two, and one that the inexperienced breeder must give intelligent heed to, is that some bitches go through the entire period of gestation without presenting a single sign of pregnancy appreciable to the ordinary observer. Of course, to a dog man the facts of the case would in all probability be known, but I shall have to confess, after years of extended experience I myself have been deceived two or three times. Never give up hope until the last gun is fired.

I think it will generally be considered a good plan, if the bitch is expected to whelp in the kennel she has been in the habit of occupying, to thoroughly clean out and wash with boiling water the box or corner she will use, to destroy all eggs and worms that may chance to be there. I also deem it a good plan to rub gently into her coat and over her breasts precipitated sulphur two or three days before the expected arrival. If the bitch is suffering from a severe case of constipation at this time, a dose of castor oil will be of service, otherwise, let her severely alone. A bitch that is in good health, properly fed, that has free access to good wholesome drinking water, can safely be left without a cathartic. Another important fact to be observed in breeding Bostons, is the suitability of certain stud dogs for particular bitches. It used to be my belief for a number of years, and I suppose many dog men today entertain the same idea, that a first class dog in every respect mated with a number of equally well bred typical bitches would produce on an average a comparatively uniform type of pups. Nothing could be further from actual results. The same dog bred, say to four females practically alike in style, size, conformation, color and markings, and from common ancestry, will give perchance in one litter two or three crackerjacks, and the other three will contain only medium pups. This same thing will occur every time the dogs are bred. This is because the bitch with the choice pups and the dog "nick," a phrase signifying that some psychological union has taken place, not understood by man, in which the best points of both dogs are reproduced in their offspring. Whenever one finds a dog eminently suited to his bitch, do not make a change, always breed to the same dog. I am perfectly cognizant of the fact that a great temptation presents itself to want to breed to a better dog, a noted prize winner probably, expecting, of course, that inasmuch as the dam did so well with a somewhat inferior dog, she must of necessity do correspondingly better with an A 1 dog. The reasoning is perfectly correct, but the result does not correspond. Very inferior pups to her previous litter by the inferior dog surprise and disgust the owner. In our kennels we have had numerous examples of this. One bitch especially, years ago, when bred to "Buster," always gave first class puppies of uniform type each litter, but the same bitch bred to some noted prize winner always gave ordinary pups. Another bitch that at the present time is practically retiring from the puppy raising business from age, when bred to Hickey's Teddy IV., always had in her litter four crackerjacks out of the seven or eight she always presented us with; when bred to any other dog (and we have tried her with several), no matter how good, never had a first class pup in the litter. Hence I repeat, if a dog "nicks" with your bitch, resulting in good pups, do not on any account ever change. Let the marriage last for life. Somewhat closely connected with this last fact is another equally important, the fact of prepotency in a stud dog, consisting of the capacity on the part of the dog to transmit his share of characteristics to his offspring in a far larger degree than is imparted by the average dog. Those who closely follow the breed will discover how certain dogs do, and have done in the past, from "Barnard's Mike" down to certain dogs of the present time, stamp the hall-mark of excellence on all the pups they sire, in a greater or less degree. Happy are those owners of dams who are aware of this important fact and take pains to use in the stud dogs of this character. I have sometimes wondered how much Barnard's Mike was worth to the breed. It will be doubtless remembered by horsemen that the great trainer, Hiram Woodruff, speaking of the importation of the thoroughbred, "Messenger," one of the founders of the American trotter, in 1788, said that "when Messenger charged down the gang-plank, in landing from the ship, the value of not less than one hundred million dollars struck our soil." He would be a very courageous man who would dare compute the worth of "Mike" or "Buster" or "Sullivan's Punch," when viewed from the same standpoint.

CHAPTER V.

REARING OF PUPPIES.

Return to Table of Contents

Assuming that the bitch has successfully whelped and all goes well, there is practically nothing to do beyond seeing that the mother is well fed, in which good meat, and where there is a good sized litter of pups, a liberal supply of milk and oatmeal gruel, is furnished. In case the mother's supply of milk is inadequate, then a foster mother must be obtained, or the pups brought up on a bottle. If a bottle, then a small one, kept scrupulously clean, with a rubber nipple that fits easily without compression. The pups must be kept perfectly warm, away from draughts, in a basket lined with flannel, and fed the first week every hour and a half day and night, every two hours the second week, and three hours in the third. I find that good, fresh cow's milk, diluted one-quarter with warm water, is the nearest approach to their natural food. After three weeks they can be fed less frequently with a spoon, and can readily be taught to lap up the milk. Where it is practical, it is always advisable to have two or more bitches whelp together, and then the pups are provided for if anything happens.

In case the bitch should lose her pups, she must be fed sparingly and her breasts should be gently rubbed with camphorated oil to prevent caking. It is not uncommon for Boston terrier pups to be born with hare-lips, in which case it is far better to put them to sleep at once, as they rarely ever live and are a deformity if they do. Be sure that the puppies' quarters have abundance of sunshine and fresh air, or they will never thrive as they should, but will be prone to disease. They are very much like plants in this respect. When the pups are four weeks old (I used to commence at five, but so many deaths have occurred in my kennels that of late I have commenced a week earlier), give them a mild vermifuge for worms. No matter if they do not show symptoms of harboring these pests, do it just the same. You will doubtless discover the reason very soon. Only those who have had experience in handling and breeding puppies are aware of their danger from worms. I know of nothing more disappointing than to go to the kennel and find the fine litter of pups that looked so promising, and on which such high hopes had been placed, with distended stomachs and the flesh literally wasted away. When this is the case do not waste a moment, administer the vermifuge. If the intestinal walls have not yet been perforated by these pests, or too great an inflammation of the alimentary canal produced, or convulsions occasioned by the impression of the worms upon the head center of the nervous system have not yet taken place, the pups, or most of them, can be saved. Hence the need of taking time by the forelock and getting rid of the worms before they get in their work. There are all kinds of worm medicines on the market, and I have tried them all. While some are all right for older pups, many of them have proven too harsh in their effects and puppies as well as worms have been destroyed. The following recipe I know will rid the little tots of their trouble without injuring them:

- Wormseed oil, sixteen drops.
- Oil of turpentine, two drops.
- Oil of anise, sixteen drops.
- Olive oil, three drachms.
- Castor oil, four drachms.

Put into a two-ounce bottle, warm slightly, shake well, and give one-half teaspoonful, floated on the same quantity of milk. If the worms do not pass away, repeat the dose the next day.

To those who would rather administer the dose in the form of a capsule, then I strongly recommend Spratts' Puppy Capsules, except when the pups are unusually small. I have just written to the Spratts people, telling them that their puppy capsules are too large for very small pups of the Boston terrier breed, and their manager has assured me he will have some made half the size. I think when the pups are about seven weeks old, when they are generally weaned, it is good, safe, precautionary measure to give them another dose of worm medicine, when we use,

- Santonine, four grains.
- Wormseed oil, twenty drops.

- Oil of turpentine, three drops.
- Olive of anise, sixteen drops.
- Olive oil, two drachms.
- Castor oil, six drachms.

Warm slightly, shake thoroughly and give one teaspoonful on an empty stomach, and I think it will be found that the worms will be eliminated. I have found it also a good plan every little while to give a teaspoonful of linseed oil to young dogs. For several years I was troubled with the loss of puppies eight or nine weeks old that had been effectually freed from worms, that seemed to gradually fade away, as it were, but an autopsy plainly revealed the cause. The mother, after eating a hearty meal, would return and vomit what she had eaten on the hay which the puppies would greedily devour. In so doing they swallowed some of the hay, which effected a lodgment in the small intestines, not being digested, until enough was collected to cause a stoppage, and the puppies consequently died. The cause being removed, we lost no more pups. As infection is always in lurk in kennels it is, I think, always advisable to give puppies that have passed the tenth week a dose of vermifuge occasionally until after the ninth month. When the kennels are kept perfectly free from fleas and other noxious insects, during the warm weather a thorough good washing once a week is of great benefit to the growing stock, and I know of no soap so good to use as the following:

- 1 lb. of Crown Soap (English harness soap).
- 1-2 ounce of mild mercurial ointment (commonly called by the chemists "blue ointment").
- 1 ounce of powdered camphor.

Mix thoroughly, and take a very small quantity and rub into the coat, thoroughly rinsing afterwards, followed by careful drying. Every day a good brushing will be found of great benefit, and when an extra luster is desired in the coat, as for the show bench, there is nothing that will do the trick as readily as to give the coat a thorough good dressing with newly ground yellow corn meal, carefully brushing out all the particles, which will leave the coat immaculately clean.

Champion Yankee Doodle Pride

Champion Dallen's Spider

Champion Mister Jack

Champion Caddy Belle

In regard to feeding the pups after weaning, it will be found an excellent plan to feed until ten weeks old four times a day, from that age until six months old, three times daily, and from that age until maturity, twice daily. I think a good drink of milk once a day excellent, and where there are enough fresh table scraps left to feed the pups, nothing better can be given. Where the number of dogs kept is too numerous to be supplied in this way, then a good meal of puppy biscuits in the morning, a good meal of meat (fresh butcher's trimmings, not too fat, bought daily) with vegetables at noon and at night well cooked oatmeal or rice with milk makes an excellent safe diet. Good, large bones with some meat on are always in order, as all dogs crave, and I think ought to have, some meat raw. Be careful not to over feed, and above all do not give the dogs sweets. When a puppy is delicate or a shy feeder, an egg beaten up in milk forms an excellent change, and good fresh beef or lamb minced up will tempt the most delicate appetite. Give the puppies a chance to get out on the fresh grass and see what Dr. Green will do for them. Above all see that they always have free access to pure, cool water.

I frequently hear numerous complaints of dog's eyes, especially pups that have been newly weaned, becoming inflamed, and in many cases small ulcers form. The same thing has occasionally happened in our kennels, and after trying practically all the eye washes on the market, sometimes without success, I applied to a friend of mine in the laboratory of the Massachusetts General Hospital and was advised by him to wash the dog's eyes two or three times a day with a ten per cent. solution of argyrol, which has been eminently successful. For slight inflammations a boracic acid wash, that any chemist will put up, will usually effect a cure.

The several forms of skin disease which cause so much disquiet to young stock, preventing rest and hindering growth, are sometimes due to faults in feeding which upset the work of the assimilative organs, and are to a great extent preventable. Not so those that are due to the presence of a parasite that burrows under the skin and produces that condition of the coat commonly known as mange. A dog may go for some considerable time unsuspected, but the sooner it is discovered and attended to the better, as it is highly contagious. The first thing to do is to take an equal amount of powdered sulphur and lard, make a paste, and rub it thoroughly into the coat of the dog and let it stay on for two days. Of course, the dog will lick off all he can, but the internal application will be good for him. At the end of the second day take the dog and give him a thorough wash with good castile soap, and after drying rub into his coat thoroughly (care being taken that none gets into the eyes or ears) crude petroleum. Let this stay on one day, and without washing take this time enough benzine and powdered sulphur to make a paste and rub in as before. It will be found that this has penetrated deeper than the lard and sulphur did and has doubtless reached the parasites. Repeat this twice, washing in between, after which give the dog a good dressing of petroleum once a day for a week, followed by a week's anointing with the benzine, and dollars to doughnuts, the dog's coat will come out all right. A good dressing to be applied occasionally afterwards, well rubbed into the skin, is composed of equal parts of castor, olive and kerosene oils, thoroughly mixed. If the hair has long been off apply the tincture of cantharides, or the sulphate of quinine to the bald spots, taking care the dog does not lick it with his tongue. These two remedies are best used in the form of an ointment, twice a day.

In regard to fleas or lice on the young stock, a good wash in not too strong a solution of any of the standard tar products is usually perfectly effectual. One other disease, and that the most deadly of all, remains to be considered, viz., distemper. This is largely contracted at the dog shows, or being brought into contact with dogs suffering from the disease. I do not believe it is ever spontaneous, and dogs kept away from infected stock will be exempt. Well do I remember my first dose of it. I had loaned a friend of mine a young dog raised by him to show, as he was trying for a prize for Druid Merk as a stud dog. The dog in question, Merk Jr., came back from the show rather depressed, and in a few days I had my entire kennel down with the disease. It was in the spring of the year, cold and damp, and I succeeded in saving just one of the young dogs and Merk Jr. After a thorough fumigation with a great quantity of sulphur I managed to get the kennels disinfected, and did not have an outbreak again for several years. A bitch sent to be bred where a case of distemper existed, unknown to me, of course, brought it to my place again, and I had the same unfortunate experience over again; fortunately this time it was in the early fall, and weather conditions being auspicious, we lost only about twenty-five per cent. of young stock. By extreme vigilance, in knowing the conditions of the kennels where bitches were sent for service, we succeeded in escaping an attack for several years, when an old bitch that had had distemper several years previously, brought back the germs in her coat from a kennel where two young dogs, just home from the Boston show, were sick with the disease. This was in the spring, the weather was wet and cold, and a loss of practically fifty per cent. ensued.

One very interesting and peculiar feature of the last attack was, that half the dogs sick were given the best medical treatment possible, with a loss of one-half; the other half were not given any medicine whatever, and the same proportion died. Of course, all had the best of care, nursing, and strict attention to diet paid.

I was very much gratified to observe that in these three attacks we have never had a dog that had a recurrence of the disease, and what is of far greater importance, have never had any after ill effect (with one solitary exception, when a bitch was left with a slight twitching of one leg) in the shape of the number of ailments that frequently follow, and in all cases after the disease had run its course the dogs seemed in a short time as vigorous as ever. This we attribute solely to the strong, vigorous constitutions the dogs possessed. A breeder who raises many dogs will have a very difficult feat to accomplish if he aspires to enter the show ring also. In our case we were convinced at the start that these two would not go together. When one considers that dogs returning from shows frequently carry the germs in their coats, and even the crates become affected, and while not suffering from the disease themselves, will readily convey it to the occupants of the kennel they come in contact with, also that the kennel man (unless a separate man has charge of infected stock exclusively) can readily carry the germs on his hands, person and clothing, it will instantly be perceived what a risk attends the combined breeding and showing. I think it pays best in the long run to keep these two branches of the business separate. The temptation to exhibit will be very strong, but before doing so, count the cost, especially if much valuable young stock is in the kennels.

In regard to the treatment of this much dreaded disease, there are a number of remedies on the market, one especially that has lately come out, viz., "Moore's Toxin," which claims to effect a cure, but having never used it can not give a personal endorsement. Whatever remedy is tried, remember that good nursing, a suitable diet, and strict hygienic measures must be given. Feed generously of raw eggs, beaten up in milk, in which a few drops of good brandy are added, every few hours, and nourishing broths and gruels may be given for a change. If the eyes are affected then the boracic acid wash; if the nose is stopped up, then a good steaming from the kettle. While the dog must have plenty of fresh air, be sure to avoid draughts. When the lungs and bronchial tubes are affected, then put flannels wrung out of hot Arabian balsam around neck and chest, and give suitable doses of cod liver oil. If the disease is principally seated in the intestines, then give once a day a teaspoonful of castor oil, and the dog should be fed with arrow root gruel, made with plenty of good milk, and a very little lean meat (beef, mutton, or chicken), once a day. When the dog is on the high road to recovery be very careful he does not get cold, or pneumonia is almost certain to ensue. Do not forget a thorough fumigation of the kennels, and all utensils, with sulphur.

CHAPTER VI.

BREEDING FOR SIZE.

Return to Table of Contents

When I joined the Boston Terrier Club in 1895, there were two classes for weight—the light weight, from 15 to 23 pounds, and the heavy weight, from 23 to 30 pounds, inclusive. This, of course, has been changed since to three classes—the light weight, 12 and not to exceed 17 pounds; middle weight class, 17 and not to exceed 22 pounds, and heavy weight, 22 and not to exceed 28 pounds and a class, for Toys, weighing under twelve pounds, has been added. The Boston terrier dog was never intended, in the writer's estimation, to be a dog to be carried in one's pocket, but such an one as the standard calls for, and which the oldest breeders have persistently and consistently bred. To my mind the ideal dog is one weighing from 15 pounds for my lady's parlor, to 20 or 25 pounds for the dog intended as a man's companion, suitable to tackle any kind of vermin, and to be an ideal watch dog in the house should any knights of the dark lantern make their nocturnal calls.

During the past few years we have had (in common, I suppose, with all large breeders), a great many orders for first class dogs, typical in every respect, weighing from 30 to 40 pounds. The constant tendency among men of wealth today is to move from the city onto country estates, where they stay the greater part of the year, and in many cases all the time. They are looking for first class watch dogs that can be kept in the house or stable, that are thoroughly reliable, that do not bring too much mud in on their coats, that do not cover the furniture with long hairs, that are vigorous enough to follow on a horseback ride, and which will not wander from home. I was in the company of a party of gentlemen the other day who had bought a number of estates in a town twenty miles from Boston, and the subject of a suitable breed of dogs for their residences was under discussion. All the fashionable breeds were gone over, some were objected to because they barked too much, others because of their propensity to rush out at teams; some that their coats were too long and they brought a great deal of mud, etc., in, and still others that their fighting disposition was too pronounced, but they all agreed that a good-sized, vigorous, good natured Boston terrier just about filled the bill. Said the nephew of Senator Henry Cabot Lodge to me last week: "Edward, I want a Boston big enough to take care of himself if anything happens, and of me also, if necessary, weighing about 35 pounds." A Boston banker, who has a large place in the country, would not take two dogs weighing under 35 pounds. Last week I received a letter from a Mr. W. B. Bogert, of the firm of Bogert, Maltby & Co., commission grain merchants, Chicago, ordering a "very heavy weight dog of kindly disposition and good blood. I can get out here any number of light weight dogs, but I do not like them. Kindly send me what you think will suit me." These are only a few sample cases, and I can say that my orders today call for more first class heavy weight dogs than for any other size. This is, of course, a comparatively new feature, but all up to date breeders will see the necessity of being able to fill this class of orders.

The small sized toys will always be in demand, as they make ideal little pets, suitable eminently for a city flat or an apartment house, to be carried by the lady in her carriage, or to accompany her in her walks, and they make first rate playmates for children. This class is by far the hardest to breed. For best results mate a bitch weighing about fifteen pounds, that comes from a numerous litter, to a twelve-pound dog that comes from small ancestry. Some of the pups are bound to be small. One important feature in the production of small pups is this: Bitches that whelp in the fall, the smallest pups are raised from, especially if the pups are fed a somewhat restricted diet, whereas puppies that are raised in the spring, that are generously fed, and have vigorous exercise in the sunshine, attain a far greater size. A great many breeders underfeed their young stock to stop growth, which I believe to be a very grave mistake. There is no question whatever it accomplishes the result wished, but at the expense of stamina and a fine, generous disposition. The pups from stock advanced in years, or from bitches excessively fat are very apt to run small, as are also the offspring of inbred parents. One very important fact in regard to breeding for large sized dogs to be considered is this: While a great many breeders always select for the production of large pups large bitches and dogs, yet experience has proven that the majority of big ones have been the offspring of medium sized dams that were bred to strong, heavy-boned dogs of substance. I bred a bitch weighing twenty pounds to a large bull terrier that weighed forty-five pounds for an experiment, and the pups, five in number, weighed at maturity from thirty-five to forty pounds, with noses and tails nearly as long as their sire's, and his color, but were very nice in their disposition, and were given away for stable dogs. Progressive up-to-date kennel men will see that they have on hand not only the three classes called for by the standard, but the fourth class, so to speak, that I have mentioned above, those weighing anywhere from thirty to forty pounds. Quite a number of breeders in the past have put in the kennel pail at birth extra large pups that they thought would mature too large to sell, but they need do so no longer. This precaution must always be taken where there are one or more of these large size puppies, viz., to look out that they do not get more than their proportionate share of the milk, or later the food, as they are very apt to crowd out the others.

Remember that the Boston terrier of whatever size will always hold his own as a companion, a dog that can be talked to and caressed, for between the dog and his owner will always be found a bond of affection and sympathetic understanding.

Prince Lutana

Champion Fosco

"Pop" Benson with Bunny II

Sir Barney Blue

CHAPTER VII.

BREEDING FOR GOOD DISPOSITION.

Return to Table of Contents

This, to my mind, is the most important feature in the breeding of the dog that demands the most careful attention. If the disposition of the dog is not all that can be desired, of what avail is superb constitution, an ideal conformation and beautiful color and markings? Better by far obtain the most pronounced mongrel that roams the street that shows a loving, generous nature if he cost his weight in gold, than take as a gift the most royally bred Boston that could not be depended upon at all times and under all circumstances to manifest a perfect disposition.

A short time ago I went to visit a noted pack of English fox hounds. One beautiful dog especially, took my eye, a strong, vigorous, noble-looking fellow, and on my asking the kennel man, a quaint old Scotchman, if he would let the dog out for me to see, he replied: "Why, certainly, Mr. Axtell, that dog is Dashwood, he is a perfect gentleman," and this is what all Boston terriers should be. Of course, I am speaking of the well bred, properly trained, blue blooded dog, not the mongrel that so often masquerades under his name. Still, as there are black sheep in every family, a dog showing an ugly, snapping, quarrelsome disposition will occasionally be met with which, to the shame of the owner, is not mercifully put out of the way and buried so deep that he can not be scratched up, but is allowed to perpetuate his or her own kind to the everlasting detriment of the breed.

How many a one has come away from a dog show utterly disgusted with perhaps one of the best looking dogs on the bench, who, after admiring its attractiveness in every detail, discovers on too near an approach to him that he possesses a snappy, vicious disposition?

I am perfectly well aware that due allowance must be made for the unnatural excitement that surrounds a dog, perhaps for the first time shown, away from all he knows, and surrounded by strange noises and faces. Yet I consider it an outrage on the public who give their time and pay their money, to subject them to any risk of being bitten by any dog, I care not of what breed it may be. At a recent show in Boston, in company with three or four gentlemen, I was admiring a very handsome looking Boston, a candidate for high honors, when his owner called out to me: "Mr. Axtell, do not go too near him or he will bite your fingers off." I replied: "You need not advise an old dog man like me; I can tell by the look of his eye what he would do if given a chance. You have no right whatever to show such a dog." Since then I went to the kennels where a noted prize winner is placed at public stud, and he showed such a vicious disposition and attempt to bite through the bars of his pen that the attendant had to cover the bars over with a blanket. Such dogs as these should be given at once a sufficient amount of chloroform and a suitable burial without mourners. If a man must keep such a brute, then a strong chain and a secure place where his owner alone can visit him is absolutely imperative.

Boston terriers, of all breeds, must possess perfect dispositions if they are to maintain their present popularity; and yet, how many unscrupulous breeders and dealers are palming off upon a confiding public dogs which, instead of being "put away" (I think that is the general term they use) should be put under so much solid mother earth that no one would suspect their interment. I know it takes considerable grit and force of character to cheerfully put to sleep a dog for which perhaps a large sum of money has been paid, that has developed an uncertain, snappy disposition, yet it pays so to do; honesty is not alone the best policy, but the only one. In my experience as a dog man I could give many personal incidents concerning the sale of vicious dogs, but for space sake one must suffice.

Last year a Chicago banker sent me an order for a dog similar in style and disposition to the one I had sold him a few years previously, to go to his niece, a young lady staying for treatment at a large sanatorium in southern Massachusetts. I replied that I had not in my kennels a large enough dog to suit, but that I knew a dealer who possessed a fairly good reputation who had, and would get him for him if he would run the chances. This was satisfactory, and I bought the dog. He was guaranteed to me as all right in every way, but I felt somewhat suspicious, as the price was very low for a dog of his style. I kept him with me for a week and saw no outs whatever about him, and practically concluded my suspicions were unfounded.

Upon taking the dog personally to the young lady in question, I told her his history as far as I knew it, and also that while I could give her the dealer's guarantee of the dog I could not of course, endorse it, but that if she cared to run the risk she could have the dog on approval as long as she wished. I said in warning that there was something about his eye that did not altogether strike my fancy, and that if he showed the least symptom of being anything but affectionate, to ship him to my kennels in Cliftondale immediately. As he was a handsome dog, with beautiful color, I could see she wanted him at once, and the dog seemed to take to her in an even greater degree. I received a letter from her in a week's time, saying how perfectly satisfactory the dog was in every way, and what a general favorite he had become with the lady patients there, several of whom would like me to get one like him for them. I need not say how pleased I was to hear this, but what was my surprise to receive a letter the next day asking me to send at once for the dog, as he had bitten the matron. You may depend that neither she nor any other of the inmates there would ever want to see a Boston again, and who would want them to? Of course I lost my money, but that is not worth mentioning. The sorrow I felt stays by me today. I sent for the dog and kept him at my kennels for five months, taking care of him myself and never letting him out of my sight, during which time he was as gentle as a kitten, until one day a young dog man came down into the yard, and the dog, for some unaccountable reason, as in the case of the matron, jumped on him and took hold of his sleeve. The man, being accustomed to dogs, was fortunately not scared. This explained the low price of the dog, and it is needless to add, he ornamented my kennels no longer. I can only state in connection with this that that dealer has sold very few dogs since. I never purchase a dog now, unless I know the man from whom I buy.

How to breed dogs possessing an ideal disposition is the all-important question, and I give the rules as followed in our kennels with complete success. Breed only from stock that you know comes from an ancestry noted for this particular feature. Many dogs are naturally of an affectionate nature, but have been made snappish by ill treatment, or teasing. This can be bred out by judicious care, but where a vicious tendency is hereditary, look out for trouble ahead. Damages for dog bites come high, and he must be either a very rich man, or a very poor one, that can afford to keep this kind of stock.

Use only thoroughly healthy stock; disease is often productive of an uneven, sullen disposition. See that the bitch especially never shows a tendency to be cross or snappy. The male dog usually controls the shape, color and markings, and the dam the constitution and disposition. Hence it is, if anything, of more importance that the female should be strong in this feature than the male, although the male, of course, should be first class also. So well known is this physiological fact that breeders of standard bred horses, particularly hunters and carriage horses, will never breed a vicious mare to a quiet stallion, and yet they are generally willing to risk breeding a quiet mare to a stallion not as good in this respect.

The education of the puppies should begin as soon as they can run around. Very much depends upon a right start. We are admonished to "train up a child in the way he should go," and this applies with equal force to the dog. Treat them with the utmost kindness, but with a firm hand. Be sure they are taught to mind when spoken to, and never fail to correct at once when necessary. A stitch in time saves many times nine. A habit once formed is hard to break. Never be harsh with them; never whip; remember that judicious kindness with firmness is far more effective with dogs, as with children. Be sure to accustom them to mingle with people and children, and introduce them as early as possible to the sights of the street, to go on ahead, and to come at your call. Prevent the pernicious habit of running and barking at teams, etc., and other dogs. The time to check these habits as aforesaid is before they become fixed. If, after all these pains, you see a dog show the slightest disposition to be vicious, then do not hesitate to send him at once by a humane transit to dog heaven. By thus continuously breeding a strain of dogs with an affectionate nature and the elimination of any that show the least deviation from the same, in a short time kennels can be established whose dogs will not only be a source of supreme satisfaction to the owner, but will be the best advertisers of said kennels wherever they go.

It will readily be admitted by all who have given the matter any consideration that a dog of an affectionate nature, whose fidelity has always been constant, and whose devotion to its owner has always under all circumstances been perfectly sincere and lasting, makes an appeal to something that is inherent in human nature. The fact of the case is that the love of such a dog is imbedded in the soul of every normal man and woman who have red blood in their veins. I think it is instinctive, and has its foundation in the fact that from the beginning of time he has ministered to man's necessities, and has accompanied him as his best friend on man's upward march to civilization and enlightenment. "There may be races of people who have never known the dog, but I very much question if, after they have made his acquaintance, they fail to appreciate his desirable qualities, and to conceive for him both esteem and affection."

Champion Lady Dainty

Champion Todd Boy

CHAPTER VIII.

BREEDING FOR A VIGOROUS CONSTITUTION.

Return to Table of Contents

I think there never was a time in the history of the breed when this particular feature needed more thoughtful, systematic and scientific attention devoted to it than now. For the past few years breeders have been straining every nerve, and leaving no stone unturned, to produce small stock, toys, in fact, and everyone realizes, who has given the question thoughtful consideration, that this line of breeding has been at the expense of the vigor, and indirectly largely of a beautiful disposition, of the dog, to say nothing of the financial loss that must inevitably ensue.

Said an old Boston terrier man (Mr. Barnard) at a recent show: "Mr. Axtell, if they keep on breeding at this rate, it won't be long before they produce a race of black and tans."

In my estimation it will not be black and tan terriers, but nothing. It will be productive of a line of bitches that are either barren, or so small that they can not possibly whelp without the aid of a "Vet." One does not have to look very far to discover numbers of men who started in the breeding of the American dog with high hopes and enthusiastic endeavors to success, who have fallen by the wayside, owing largely to the fact that proper attention was not paid to the selection of suitable breeding stock, especially the matrons. Said a man to me last year: "Much as I love the dog, and crazy as I am to raise some good pups, I have given up for all time trying to breed Boston terriers. I have lost eight bitches in succession whelping." We have all of us "been there" and quite a number of us "many a time."

In order to obtain strong, vigorous puppies that will live and develop into dogs that will be noted for vigorous constitutions, we shall simply, and in language that can be readily understood by the novice as well as the established breeder, lay down the rules that a quarter of a century has demonstrated to be the correct ones for the attainment of the same as used in our kennels. As all puppies that leave our place are sold with the guarantee of reaching maturity (unless shown, when we take no risks whatever in regard to distemper, mange, etc.), it will readily be seen that they must have a first class start, and must of necessity be the progeny of stock possessing first class vigor and the quality of being able to transmit the same to their offspring. An ounce of experience is worth many tons of theory, and it is, then, with pleasure we give the system pursued by us, feeling certain that the same measure of success will attend others that will take the necessary pains to attain the same, and they will be spared the many pitfalls and mistakes that have necessarily been ours before we acquired our present knowledge. It has been for a number of years (starting as we did when the breed was in its infancy, and only the intense love of the dog, coupled with an extensive leisure, which enabled us to devote a great deal of attention to important and scientific experiments, have enabled us to arrive where we are), an uphill road, the breeding problems have had to be solved at the outlay of brains, patience and considerable money. Unlike any established breed, there was practically no data to fall back on, no books of instruction to follow, but if the pioneer work has been arduous the results obtained have far outbalanced it, and the dog today stands as a monument to all the faithful, conscientious and determined body of men who would never acknowledge defeat, but who, in spite of all discouragements from all quarters, and from many where it should have been least expected, have pressed forward until they find the object of their unfailing endeavors the supreme favorite in dogdom the continent over.

In the first place, in the attainment of vigorous puppies, we state the bitches selected are of primary importance, in our view, as already stated, far more so than the sire. For best results we choose a bitch weighing from fifteen to twenty-five pounds. If they happen to weigh over this we do not consider it any detriment whatever, rather otherwise. Always select said matrons from litters that have been large, bred from strong, vigorous stock, thoroughly matured, and that have been bred by reliable (we speak advisedly) men for several generations if possible. If one can, obtain from kennels that while perfectly comfortable, have not been supplied with artificial heat. There is more in this than appears on the surface. Dogs that have been coddled and brought up around a stove rarely have stamina and vitality enough to enable them to live the number of years they are entitled to, and fall a ready victim to the first serious trouble, whether distemper, or the many and one ills that beset their path. Intelligent breeders of all kinds of stock today recognize the value of fresh air and unlimited sunshine, and if best results are to be obtained these two things are imperative.

I was very much interested in the prize herd of Hereford cattle owned by Mr. Joseph Rowlands, near Worcester, England, and conceded by experts to be the best in that country, and to learn that for a number of years the herd (over one hundred in number) have been kept in the open, the cows being placed in the barn for a few days at calving, and that the prize winning bull that heads the herd, "Tumbler," is sixteen years old, and still used, and it is stated by Mr. Rowlands is producing as good stock today as ever. The significant fact about this herd is, they are and have been perfectly free from tuberculosis. Another herd of Jerseys (although not prize winners) are kept near there, under precisely the same conditions with similar results. A breeder of prize winning Belgian hares has kept these for a number of years without artificial heat, with the best of results with freedom from disease, and the attainment of strong, robust constitutions. When puppies are four months old (in the winter time) they should be placed in well built kennels, without artificial heat. (Of course, this does not apply to a colder latitude than Massachusetts.)

The reason for choosing bitches that come from dams noted for their large litters is this: the chances are (if the dog bred to comes from a similar litter) that they will inherit the propensity to give birth to large litters themselves, and the pups will necessarily be smaller than when only one or two pups are born. The bitch that has but that number runs an awful risk, especially if she has been well fed. The pups will be large and the dam has great difficulty in whelping.

If toy bitches are bred, look out for breakers ahead; only a very small per cent. live to play with their little ones. A toy bitch, bred to a toy dog, will frequently have but one pup, and that quite a large one in proportion to the size of parents. When a toy bitch is bred, attend carefully to these three things. See that the dog used is small in himself, comes from small stock, and does not possess too large a head. Secondly, be sure the bitch is kept in rather poor condition, in other words, not too fat; and thirdly, and this is the most important of all, see that she has all the natural exercise she can be induced to take. These conditions strictly and faithfully adhered to may result in success.

In the next place, the consideration of the dog to be used is in order. Whether he be a first prize winner or an equally good dog that has never been shown (and the proportion of the best raised dogs that appear on the bench is very small) insist on the following rules:

Be sure that the dog is typical with first class constitution, vigorous, and possessing an ideal disposition, and what is of the utmost importance, that he comes from a line of ancestry eminently noted for these characteristics. Breed to no other, though he were a winner of a thousand first prizes. I prefer a symmetrical dog weighing from sixteen to twenty pounds, rather finer in his make-up than the bitch, and possessing the indefinable quality of style, and evidences in his make-up courage and a fine, open, generous temperament. Do not breed to a dog that is overworked in the stud, kept on a board floor chained up in a kennel or barn, and never given a chance to properly exercise. If you do the chances are that one of three things will happen: the bitch will not be in whelp (the most likely result) the pups, or some of them will be born dead, and one runs an awful risk of the bitch dying, or, if alive at birth, a very small per cent. only of the pups will live to reach maturity. I think Boston terriers are particularly susceptible to worms or distemper, and it is absolutely imperative that they should not be handicapped at the onset.

One other very important factor is natural exercise for the bitch. Unless one is willing to take the necessary pains to give her this, give up all expectation of ever succeeding in raising puppies.

Champion Willowbrook Glory

Squantum Punch

Tony Ringmaster

Someone asked a noted critic whom he considered the best singer he had ever heard, and he answered, "Patti." In being asked who came next, he replied, "Patti;" and on being questioned who was his third choice, gave the same answer. Were I asked the three most important essentials for the success of the brood bitch, I should say, "Exercise, exercise, exercise." By this I do not mean leading with a chain, running behind a horse or team, but the natural exercise a bitch will take if left to her own devices. Nature has provided an infallible monitor to direct the dog the best amount to take, and when to take it. One of the best bitches I ever possessed was one weighing fourteen pounds by the original Tony Boy (one of the best little dogs that ever lived) out of a bitch by Torrey's Ned, by A. Goode's Ned. Her name was Lottie, and she had thirteen litters and raised over ninety per cent. Those who have read that interesting little book on the "Boston Terrier," by the late Dr. Mott, will readily recall the genial Doctor speaking of the first Boston he ever owned, named "Muggy Dee," and how intelligent he was, and what a number of tricks the Doctor taught him, will be interested to know that Lottie was his great-grandmother, and she was equally intelligent. We had several bitches by the celebrated Mr. Mullen's "Boxer" out of her, (this is going back to ancient history), one of which, "Brownie," was, to my fancy, the nicest dog we ever had. She, with the rest of the litter, had the run of several hundred acres, and many times I did not see them for days together. They went in and out of the hayloft at pleasure, and spent the greater part of their time hunting and digging out skunks and woodchucks which were quite thick in the woods back of us at that time. I remember the first time Brownie was bred to that king of sires, "Buster," owned by Alex. Goode (than whom a more loyal Boston terrier man never lived), and I was rather anxious to see the litter when it arrived, as from the mating I expected crackerjacks. I had not seen her or her mother for two or three days, but the time for whelping having arrived, was keeping a close watch on the stable. About dusk she came in with Lottie, and in a short time gave birth to four of the most vigorous, perfectly formed little tots I had ever seen. Each one proved to be good enough to show, although only one was sold to an exhibitor, Mr. G. Rawson, the rest going into private hands. "Druid Pero" was shown in New York in 1898, taking first prize and silver cup for best in his class, but I think his brother, "Caddie," beat him, his owner, a Boston banker, being offered a number of times ten times the sum he paid for him.

The day after Brownie whelped she and her mother went off for an hour or so, and they finished digging out Mr. Skunk (which the attention to her maternal duties necessitated a postponement of), the old dog dragging him home in triumph. I attribute the success these dogs, in common with the rest of the bitches in the kennels who had similar advantages, had in whelping and the rearing of their young to the fact that they always had unlimited natural exercise. I can enumerate scores of cases similar to these attended with equally good results, if space permitted.

In regard to mating, one service, if properly performed, is usually enough, if the bitch is ready to take the dog. If a bitch should fail to be in whelp I should advise the next time she comes in season two or even three visits to the dog, and where convenient I should suggest a different dog this time. In case this time these services were unsuccessful, then I should suggest the course that breeders of thoroughbred horses pursue, viz., to let the female run with the male for three or four days together. There are many things connected with breeding that we do not understand, and frequently going back to nature, as in this case, is productive of results when all else fails.

One very important factor in the production of strong, rugged pups that live, is good feeding. Do not imagine that feeding dog biscuits to the bitch in whelp will give good results, it will not; she needs meat and vegetables once a day. Biscuits are all right as a supplementary food, but that is all. Meat is the natural food for a dog, and it is a wise kennel man that can improve on nature. Be sure the meat is free from taint, especially at this time and when the bitch is nursing pups. The gastric juice of a dog's stomach is a great germicide, but there is a limit.

Be certain the dogs have a plentiful supply of good, pure water. This is of far more importance than many people imagine.

Do not administer drugs of any description to your dogs, except in the case of a good vermifuge, if they are harboring worms, and a proper dose of castor oil if constipated. If the dog at any time is sick, consult a good veterinary accustomed to dogs, not one who has practiced entirely on horses or cows. If a bitch, at the time of whelping, is much distressed and can not proceed, get a veterinary and get him quick. When the pups arrive, if all is well and they are able to nurse, let them severely alone. If they are very weak they will have to be assisted to suckle—do not delay attention in this case. Be sure the box the bitch whelped in is large enough for her to turn around in, and do not use any material in the nest that the pups can get entangled with. My advice to breeders is, if the bitch is fully formed and grown to her full proportions, to breed the first time she comes in season. She will have an easier time whelping than when she is older. If delicate or immature, delay breeding till the next time. Do not use a dog in the stud until he is a year and a half old for best results; they will, of course, sire pups at a year or younger, but better wait. To those people who live in the city, or where a kennel can not be established for want of adequate room to give the dogs the necessary exercise, an excellent plan to follow is one adopted by an acquaintance of mine, and followed by him for a number of years with a good measure of success. He owns one or two good stud dogs that he keeps at his home, and he has put out on different farms, within a radius of ten miles of Boston, one bitch at each place, and pays the farmer (who is only too glad to have this source of income at the outlay of so little trouble and expense) one hundred dollars for each litter of pups the bitch has, the farmer to deliver the pups when required, usually when three months old. The farmer brings in the bitch to be bred, and the owner has no further trouble. The pups, when delivered, are usually in the pink of condition and are, in a great measure, house broken, and their manners to a certain extent cultivated. He has no trouble whatever with pups when ordered, as he simply sends the address of customers and the farmer ships them. This, to me, is a very uninteresting and somewhat mercenary way of doing business, as one misses all the charm of breeding and the bringing up of the little tots, to many of us the most delightful part of the business. To those breeders who have newly started in, do not get discouraged if success does not immediately crown your efforts; remember, if Boston terriers could be raised as easily as other dogs, the prices would immediately drop to the others' level.

Goode's Buster

Champion Whisper

Champion Druid Vixen

Champion Remlik Bonnie

CHAPTER IX.

BREEDING FOR COLOR AND MARKINGS.

Return to Table of Contents

Every one who has a Boston terrier for sale knows that a handsome seal or mahogany brindle with correct markings, with plenty of luster in the coat, provided all other things are equal, sells more readily at a far higher price than any other. When one considers the number of points given in the standard for this particular feature, and the very important factor it occupies in the sale of the dog, too much attention cannot be given by breeders for the attainment of this desideratum. I am, of course, thoroughly in sympathy with the absolute justice that should always prevail in the show ring in the consideration of the place color and markings occupy in scoring a candidate for awards. Twelve points are allowed in the standard for these, and any dog, I care not whether it be "black, white, gray, or grizzled," that scored thirteen points over the most perfectly marked dog, should be awarded the prize. But be it ever remembered that the show ring and the selling of a dog are two separate and distinct propositions. In the writer's opinion and experience a wide gulf opens up between a perfect white or black dog comporting absolutely to the standard, and one of desirable color and markings that is off a number of points. I have always found a white, black, mouse, or liver colored dog, I care not how good in every other respect, almost impossible to get rid of at any decent price. People simply would not take them. Perhaps my experience has run counter to others. I trust it may have done so, but candor compels me to make this statement.

I find that this condition of things is somewhat misleading, especially to beginners in the breed. They have seen the awards made in the shows (with absolute justice, as already stated), and have naturally inferred that in consequence of this, breeding for desirable colors was not of paramount importance after all. Only a month or two ago an article appeared in a charming little dog magazine, written evidently by an amateur, on this question of color and markings. He had visited the Boston Terrier Club show last November, and speaking of seal brindles, said: "If this color is so very desirable it seems strange that so few were seen, and that so many of the leading terriers were black and white, and some white entirely," then follows his deduction, viz., "the tendency evidently is that color is immaterial with the best judges, so that a breeder is foolish to waste his time on side issues which are not material." I can only state in passing that if he had a number of dogs on hand that were of the colors he specifies, "black and white, and some white entirely," it would doubtless "seem strange" to him why they persisted in remaining on his hands as if he had given each one an extra bath in Le Page's liquid glue. Pitfalls beset the path of the beginner and this book is written largely to avoid them. When one reads or hears the statement made that color and markings are of secondary consideration or even less, take warning. The reader's pardon will now have to be craved for the apparent egotism evidenced by the writer in speaking of himself in a way that only indirectly concerns canine matters, but which has a bearing on this very important question of color, and partially, at least, explains why this particular feature of the breeding of the Boston terrier has appealed to him so prominently. My father was a wholesale merchant in straw goods, and had extensive dye works and bleacheries where the straw, silk and cotton braids were colored. As a youngster I used to take great delight in watching the dyers and bleachers preparing their different colors and shades, etc., and was anxious to see the results obtained by the different chemical combinations. When a young man, while studying animal physiology under the direction of the eminent scientist, Professor Huxley, whose diploma I value most highly, I made a number of extended scientific experiments in color breeding in poultry and rabbits, so that when I took up breeding Boston terriers later in life this feature particularly attracted me. I was "predisposed," as a physician says of a case where the infection is certain, hence I offer no apology whatever for the assertion that this chapter is scientifically correct in the rules laid down for the breeding to attain desirable shades and markings.

When we first commenced breeding Bostons in 1885, the prevailing shades were a rather light golden brindle (often a yellow), and mahogany brindles, and quite a considerable number had a great deal of white. Then three shades were debarred, viz., black, mouse and liver, and although years after the Boston Terrier Club removed this embargo, they still remain very undesirable colors.

The rich mahogany brindle next became the fashionable color (and personally I consider it the most beautiful shade), and Mr. A. Goode with Champion "Monte" and Mr. Rawson with the beautiful pair, "Druid Merke" and "Vixen," set the pace and every one followed. A few years later Messrs. Phelps and Davis (who, with the above mentioned gentlemen, were true friends of the breed), sold a handsome pair of seal brindles, Chs. "Commissioner II." and "Topsy," to Mr. Borden of New York, and confirmed, if not established, the fashion for that color in that city. I think that all people will agree, from all parts of the country, that New York sets the style for practically everything, from my lady's headgear to the pattern of her equipages, and the edict from that city has decreed that the correct color in Boston terriers is a rich seal brindle, with white markings, with plenty of luster to it, and all sections of the continent promptly say amen!

I have taken the pains to look up a number of orders that we have recently received, which include (not enumerating those received from the New England States, or New York), three from Portland, Oregon, one from California, one from St. Louis, one from Mexico, four from Canada, two from Chicago, and one from Texas, and with the exception of two who wished to replace dogs bought of us ten or twelve years previously, they practically all wanted seal brindles.

These orders were nearly all from bankers and brokers, men who are supposed to be en rapport with the dictates of fashion. It goes without saying that what a public taste demands, every effort will be made to attain the same, and breeders will strive their utmost to produce this shade. Many who do not understand scientific matings to obtain these desirable colors have fallen into a very natural mistake in so doing. In regard to the mahogany brindles they say, why not breed continuously together rich mahogany sires and dams, and then we shall always have the brindles we desire. "Like produces like" is a truism often quoted, but there are exceptions, and Boston terrier breeding furnishes an important one. A very few years of breeding this way will give a brown, solid color, without a particle of brindle, or even worse, a buckskin. If the foundation stock is a lighter brindle to start, the result will be a mouse color. The proper course to pursue is to take a golden brindle bitch that comes from a family noted for that shade, and mate her with a dark mahogany brindle dog that comes from an ancestry possessed of that color. The bitch from this mating can be bred to dark mahogany brindles, and the females from this last mating bred again to dark mahogany males, but now a change is necessary. The maxim, "twice in and once out," applies here. The last bred bitches should be bred this time to a golden brindle dog, and same process repeated, that is, the bitches from this last union and their daughters can be bred to dark mahogany brindle dogs, when the golden brindle sire comes in play again. This can be repeated indefinitely. A rule in color breeding to be observed is this: that the male largely influences the color of the pups. If darker colors are desired, use a darker male than the female. If lighter shades are desired, use a lighter colored male.

If a tiger brindle is wanted, take a gray brindle bitch and mate to a dark mahogany dog. Steel and gray brindles are in so little demand and are so easy to produce that we shall not notice them.

In regard to seal brindles. A great many breeders who do not understand proper breeding to obtain them have fallen into the same pit as the others. In their desire to obtain the dark seal brindles they have mated very dark dogs to equally dark bitches, which has resulted in a few generations in producing dogs absolutely black in color, with coats that look as if they had been steeped in a pail of ink. A visit to any of the leading shows of late will reveal the fact that quite a number of candidates for bench honors are not real brindle, except possibly on the under side of the body, or perchance a slight shading on the legs. A considerable number are perfectly black, and are called by courtesy black brindles. As well call the ace of spades by the same name. A serious feature in connection with this is, that the longer this line of breeding is persisted in, the harder will be the task to breed away. In fact, in my estimation it will be as difficult as the elimination of white. One important fact in connection here is that black color is more pronounced from white stock than from brindle. I recently went into the kennels of a man who has started a comparatively short time ago, and who has been most energetic in his endeavors to produce a line of dark seal brindles, and who is much perplexed because he has a lot of stock on hand, while first rate in every other respect, are with coats as black as crows and not worth ten dollars apiece. He seemed very much surprised when I told him his mistake, but grateful to be shown a way out of his difficulty. A visit to another kennel not far from the last revealed the fact that the owner was advertising and sending largely to the West what he called black brindles, but as devoid of brindle as a frog is of feathers. His case was rather amusing, as he honestly believed that because the dog was a Boston terrier its color of necessity must be a brindle. He reminded me a good deal of a man who started a dog store in Boston a number of years ago who advertised in his windows a Boston terrier for sale cheap. Upon stepping in to see the dog all that presented itself to view was a dog, a cross between a fox and bull terrier. When the man was told of this, he made this amusing reply: "The dog was born in Boston, and he is a terrier. Why is he not a Boston terrier?" Upon telling him that according to his reasoning if the dog had been born in New York city he would be a New York terrier he smiled. Fortunately I had "Druid Pero" with me and said: "Here is a dog bred in my kennels at Cliftondale, Mass., that was a first prize winner at the last New York show, and yet he is a Boston terrier." After looking Pero carefully over he exclaimed: "Well, by gosh, they don't look much like brothers, but I guess some greenhorn will come along who will give me twenty-five dollars for him," and on inquiring a little later was told the green gentleman had called and bought the dog.

How to breed the dogs so that the brindle will not become too dark, with the bright reddish sheen that sparkles in the sun, is the important question, and I am surprised at the ignorance displayed by kennel men that one would naturally suppose would have made the necessary scientific experiments to obtain this desirable shading. Only a short time ago a doctor, a friend of mine, told me he had just started a kennel of Bostons, buying several bitches at a bargain on account of their being black in color, and that he proposed breeding them to a white dog to get puppies of a desirable brindle. He seemed quite surprised when told the only shades he could reasonably expect would be black, white and splashed, all equally undesirable.

The system adopted in our kennels some years ago to obtain seal brindles with correct markings and the desirable luster and reddish sheen to the coat is as follows:

We take a rich red, or light mahogany bitch, with perfect markings, that comes from a family noted for the brilliancy of their color, and without white in the pedigrees for a number of generations, and mate her always to a dark seal brindle dog with an ancestry back of him noted for the same color. The pups from these matings will come practically seventy-five per cent. medium seal brindles. We now take the females that approximate the nearest in shade to their mother, and mate them to a dark seal brindle dog always. The bitches that are the result of this union are always bred to a dark seal brindle dog. The females that come from the last union are bred to a medium seal brindle dog, but now comes the time to introduce a mahogany brindle dog as a sire next time, for if these last bitches were mated to a seal brindle dog a large per cent. of the pups would come too dark or even black. This system is used indefinitely and desirable seal brindles with white markings can thus be always obtained. To the best of my recollection we have had but one black dog in twenty years. We have demonstrated, we trust, so that all may understand how golden, mahogany, and seal brindles are obtained, and how they may be bred for all time without losing the brindle so essential, and we now pass on to the consideration of a far harder problem, the obtaining of the rich seal brindles from all undesirable colors, and we present to all interested in this important, and practically unknown and misunderstood, problem the result of a number of years extended and scientific experiments which, we confess, were disheartening and unproductive for a long time, but which ultimately resulted in success, the following rules to be observed, known as "The St. Botolph Color Chart."

In presenting this we are fully aware that as far as we know this is the only scientific system evolved up to date, also that there are a number of breeders of the American dog who maintain that this is an absolute impossibility, that breeding for color is as absurd as it is impractical, but we can assure these honest doubters that we have blazed a trail, and all they now have to do is simply to follow instructions and success will crown their efforts.

We will enumerate the following colors in the order of their resistance, so to speak:

No. 1. White. This color, theoretically a combination of red, green and violet will be found the hardest to eliminate, as the shade desired will have to be worked in, so to speak, and it will take several generations before a seal brindle with perfect markings that can be depended upon to always reproduce itself can be obtained. Starting with a white bitch (always remember that the shades desired must be possessed by the dog), we breed her always to a golden brindle dog. The bitches (those most resembling the sire in color being selected) from these two are mated to a dark mahogany brindle dog, and the females from this last union are mated to a dark seal brindle dog. It will readily be observed that we have bred into the white color, golden, mahogany and seal brindle and this admixture of color will give practically over ninety per cent. of desirable brindles. Always see that the sires used are perfectly marked, from ancestry possessing the same correct markings. This is absolutely imperative, where the stock to be improved is worked upon is white.

No. 2. Black. This color is the opposite of white, inasmuch as there is an excess of pigment, which in this case will have to be worked out. Breed the black bitch to a red brindle dog (with the same conditions regarding his ancestry). The females from these matings bred always to a dark mahogany brindle dog. The females from the last matings breed to a medium seal brindle dog with a very glossy coat, and the result of these last matings will be good seal brindles. If any bitches should occasionally come black, breed always to a golden brindle dog. No other shade will do the trick.

No. 3. Gray brindle. This is practically a dead color, but easy to work out. Breed first to a golden brindle dog. The females from this union breed to a rich mahogany brindle, and the bitches from this last litter breed to a seal brindle dog.

No. 4. Buckskin. Breed bitch to golden brindle dog; the females from this union to a red brindle dog (if unobtainable, use mahogany brindle dog, but this is not so effective), and the females from last union breed to a seal brindle dog.

No. 5. Liver. This is a great deal like the last, but a little harder to manipulate. Breed first to a golden brindle dog. The females from this union breed to a seal brindle. The bitches from this union breed to mahogany brindle dog with black bars running through the coat, and the females from last mating breed to seal brindles.

No. 6. Mouse color. Use same process as for gray brindles.

No. 7. Yellow. A very undesirable shade, but easy to eliminate. Breed to mahogany brindle dog as dark as can be obtained, and bitches from this mating breed to a seal brindle dog.

No. 8. Steel and tiger brindles I class together, as the process is the same and results are easy. Breed first to a red brindle dog; bitches from this union to a dark mahogany brindle, and then use seal brindle dog on bitch from last mating.

No. 9. Red brindle. No skill is required here. Breed first to mahogany brindles, and bitches from this union to seal brindles.

We have now enumerated practically all the less desirable shades, but let me observe in passing, in the process of color breeding that the law of atavism, or "throwing back," often asserts itself, and we shall see colors belonging to a far-off ancestry occasionally presenting themselves in all these matings. Once in a while a dog will be found that no matter what color bitches he may be mated with, he will mark a certain number of the litter with the peculiar color or markings of some remote ancestor. Just a case apropos of this will suffice. We used in our kennels a dog of perfect markings, coming from an immediate ancestry of perfectly marked dogs, and mated him with quite a number of absolutely perfectly marked bitches that we had bred for a great number of years that had before that had perfectly marked pups, and every bitch, no matter how bred, had over fifty per cent. of white headed pups. We saw the pups in other places sired by this dog, no matter where bred, similarly marked. We found his grandmother was a white headed dog, and this dog inherited this feature in his blood, and passed it on to posterity. The minute a stud dog, perfect in himself, is prepotent to impress upon his offspring a defect in his ancestry, discard him at once. I have often been amused to see how frequently this law of atavism is either misunderstood or ignored. Only recently I have seen a number of letters in a leading dog magazine, in which several people who apparently ought to know better, were accusing litters of bulldog pups as being of impure blood because there were one or two black pups amongst them. They must, of course, have been conversant with the fact that bulldogs years ago frequently came of that color, and failed to reason that in consequence of this, pups of that shade are liable once in a while to occur. It is always a safe rule in color breeding to discard as a stud a dog, no matter how brilliant his coat may be, who persistently sires pups whose colors are indistinct and run together, as it were.

Champion Boylston Reina

Champion Roxie

Peter's Little Boy and Ch. Trimont Roman

Champion Lord Derby

Remember, in closing this chapter, that as "eternal vigilance is the price of liberty," so the eternal admixtures of colors is the price of rich brindles. If one has the time the works of an Austrian monk named Mendel are of great interest as bearing somewhat on this subject, and the two English naturalists, Messrs. Everett and J. G. Millais, whose writings contain the result of extensive scientific experiments on dogs and game birds, are of absorbing interest also.

CHAPTER X.

SALES.

Return to Table of Contents

Every person who has bred Bostons for any length of time knows that a good dog sells himself. I do not imagine there is practically any part of this great country where a typical dog, of proper color and markings and all right in every respect, fails to meet a prospective buyer, and yet, of course, there are certain places where an A 1 dog, like an ideal saddle or carriage horse meets with a readier sale, at a far greater price than others. New York city, in particular, and all the larger cities of the country where there are large accumulations of wealth, offer the best markets for the greatest numbers of this aristocratic member of the dog fraternity, and from my own personal knowledge the larger cities of the countries adjacent to the United States furnish nearly as good a market, at a somewhat reduced price. Were the quarantines removed in the mother country, which England no doubt has found absolutely necessary, it would not surprise me in the least to see an unprecedented demand for the Boston at very high prices, and I am going to make a prediction that on the continent of Europe it will not be long before the American dog will follow the trotting horse, and will work his way eastward, until jealous China and strange Japan will be as enamoured with him as we are, and his devotees at the Antipodes will be wondering where he got his little screw tail, and why that sweet, serene expression on his face, like the "Quaker Oat smile," never comes off. This to a person who knows not the Boston may seem extravagant praise, but to all such we simply say: Get one, and then see if you are not ready to exclaim with the Queen of Sheba, when visiting King Solomon and being shown his treasures: "Behold, the half was not told me!" Perhaps the system of sales that has always been followed by us may be of interest to many engaged in the breeding of the dog, and while we do not hold a patent on the same, or even suggest its adoption by others, must confess it has worked with entire satisfaction in our case, and we have never once failed to receive the purchase money. We must say in explanation that our customers practically are all bankers and brokers, and that our dogs have never been sold by advertising or being exhibited at shows, but by being recommended by one man to another, starting many years ago by the first sale to a Boston banker, then to several members of his firm, going from Boston to their correspondents in other cities, until the orders come in from everywhere. We had three orders from as many countries in one mail last week. I merely mention this to show how the demand for the dog has grown. When we commenced to sell dogs we adopted the following plan, which we conceived to be just and equitable alike to buyer and seller: When a dog is ordered we send on one which we believe will fill the bill, accurately describing the dog, stating age, pedigree, etc., and stating that when the customer is perfectly satisfied with the dog (as long a trial being given as may be wished) in every respect, a check will be accepted, and not before. Should the dog at any time prove unsatisfactory in any way, the purchase money will be cheerfully refunded, or a dog of equal value will be sent in exchange. In the case of a bitch that fails to become a good breeder, the same plan, of course, is followed. In regard to the sale of puppies, we guarantee them (barring accidents, and the showing of them, when owner assumes risks) to reach maturity, and in case they do not, refund purchase money, or send on another puppy of equal value.

Of course, where the buyer is not known, or personally recommended, then the seller has to adopt entirely different methods. Still, I see no reason why an honest man who has a Boston, or any other dog, for sale, or, in fact, any article of merchandise, should not be willing to send on the same to any honest buyer. This is on the assumption, of course, that both parties are honorable men. To the seller I advise the purchase money being received before the dog is shipped, and express charges guaranteed, if the buyer is not known or unable to supply absolutely reliable references. Decline to receive any order where the object sought is to obtain a dog to use to breed to a bitch, or several, as the case may be, and then be returned as unsatisfactory. We have had no experience in this line, but are informed it has frequently been done. If such a customer presents himself, simply tell him he can inspect the dog or have an expert do so for him if too far away to come, but that when the deal is closed and the money paid that under no conditions whatever can the dog be returned. In regard to the seller shipping the dog to its destination, we will say that we think he will run practically no risk in so doing. If the dog is all right in every way it is dollars to doughnuts that he will arrive in perfect condition. We can say that in over twenty years' shipments of dogs to all parts of the country and beyond we have never had a dog die en route, lost, exchanged, or stolen. I think the express companies of this country, Canada, Mexico, and beyond, are to be highly commended for the excellent care they take of the dogs committed to their charge, neither do I think the express charges are ever excessive, when one considers the value of the dogs carried.

We will now consider the case of the buyer, assuming, of course, he is known or capable of presenting suitable references. We always advise him to deal with kennels or dealers of established reputations. Run no chances with any other unless you desire to be "trimmed." Pray do not be misled by glowing advertisements (stating that they have the largest kennels on earth) in every paper that does not know them. I have investigated quite a number of these so-called kennels and found they usually consisted of an old box stall in a cheap stable, or a room over an equally cheap barroom, and their stock in trade consisted of two or three mutts.

Be very suspicious of any man who advertises that he has dogs for sale that can win in fast company for fifty or a hundred dollars, or A 1 bitches in whelp to noted dogs for the same price. Any man who possesses these kinds of dogs does not have to advertise their sale. There are plenty of people here in Boston only too glad to buy this kind of stock at three or four times this price.

I attended the last show in Boston with a number of orders in my pocket, but failed to discover any dogs I picked out possessing the quality described at anything less than a good stiff price, for Boston terriers with the "hall mark" of quality have been, are, and, I believe, always will be, as staple in value as diamonds.

The number of letters we have received from all over the country, particularly from the West, complaining of the skin games played upon them by fake kennels and dealers, would make an angel weep, and make one almost regret that one ever knew a Boston. If the same ingenuity, skill and patience employed in the getting up of these fake advertisements had been devoted to the breeding of the dog, this class of advertising gentry (?) would have produced something fit to sell. It is stated on the best of authority that in some cases nothing was shipped for money received.

In spite of this vast number of unscrupulous breeders and dealers scattered abroad, I think the chances for reliable kennels was never so good as now in the history of the breed. Cream will always rise, and right dealing, whether in dogs or diamonds, will ever meet with their just returns. Remember that one never forgets being "taken in" in a horse trade, and when, instead of a horse a dog is involved, I think one never forgives as well. To that number of persons who, in their daily walks of life are fairly honest, but who, when it comes to a trade in dogs are apt to lose that fine sense of justice that should characterize all transactions, we would say with Shakespeare: "To thine own self be true. Thou canst not then be false to any man." Yea, we would repeat the command of a greater than Shakespeare, to whom, I trust, we all pay reverence, when He lays down for us all the Golden Rule: "Whatsoever ye would that men would do to you, do ye even so to them."

To go back to the responsible buyer who is in the market for a good dog, we say: Send your orders to responsible men, with said dogs to sell, stating exactly what you want, and the price you desire to pay, agreeing to send a check just as soon as dogs prove satisfactory, assuming, of course, express charges. Reputable dealers and breeders are looking for just such customers.

To all breeders and dealers who have not an established reputation, would say: Advertise accurately what you have for sale in first class reliable papers and magazines. In regard to prices, the following scale, adopted by us many years ago, and which we have never seen since any reason to change, is practically as follows:

For pups from two to three months old, from fifty to seventy-five dollars. When six months old, from seventy-five to a hundred: From six months to maturity, from one hundred to two hundred. These prices are, of course, for the ordinary all-around good dogs. With dogs that approximate perfection, and which only come in the same proportion as giants and dwarfs do in the human race (I believe the proportion is one in five thousand), and the advent of which would surprise the average kennel man as much as if the President had sent him a special invitation to dine with him at the White House, the price is problematical, and is negotiated solely by the demand for such a wonder by a comparatively few buyers.

I think Boston terriers as a breed occupy the same position amongst dogs as the hunter and carriage horse does amongst horses. Each are more or less a luxury. A well matched pair of horses of good all-round action, of desirable color and perfect manners and suitable age will sell in the Eastern cities (I am not sufficiently acquainted with the other sections of the country to know values there) at from eight hundred to two thousand dollars, but with a pair of carriage horses able to win on the tan bark, the price will be regulated by the comparatively few people who have sufficient money to spare to purchase this fashionable luxury, and ten times the amount paid for the first mentioned pair would be a reasonable price to pay for the prize winners. I think the winners of the blue in the Bostons would fetch a relative sum.

The important factor of the cost of production in the case of the dog necessarily enters into the selling price. Good Bostons are as hard to raise as first class hunters, and a correspondingly large sum has to be obtained to meet expenses, to say nothing of profit, but in the writer's experience the best dog or horse sells the readiest. Do not be misled by the remark "that a dog is worth all he will bring." Generally speaking, this is sound logic, but not always. Many dogs have been sold for very little by people not cognizant of their value, but this in no way changed the intrinsic worth of the dog. On the other hand, many dogs have been disposed of at many times their real value, but this transaction did not enhance their worth in the slightest degree. A gold dollar is worth one hundred cents whether changed for fifty cents or five hundred. An article of intrinsic value never changes. Our advice to all who have dogs for sale (or any other article, in fact), ask what you know is a good, honest, fair value, and although you may not sell the dog today, remember that there are other days to follow. What I am going to add now I know a great many dealers and breeders will laugh at and declare me a fit subject for an alienist to work on, but it is fundamentally true just the same, and is this: Never ask or take for a dog more than you know (not guess) the dog is worth. This is nothing but ordinary, common everyday justice that every man has every right to demand of his fellow man, and every man that is a gentleman will recognize the truth and force of.

I was reading a novel this summer, and one statement amongst a great many good ones impressed me. It stated "that all men were divided into two classes: those that behaved themselves, and those who did not." We all know that society has divided men into many classes, but I think any thoughtful man will confess, in the last analysis, that the novelist's classification was the correct one. I need not apply the moral.

It will be somewhat of a temptation to resist taking what a party, liberally supplied with this world's goods, will frequently in their ignorance offer for a dog that appeals to them, but which the owner knows perfectly well is not worth the price offered. If he belongs to the class that behaves themselves he will tell the prospective buyer what the dog is intrinsically worth, and point out the reasons why he is not worth more. You may depend that you have not only obtained a customer for life, but one that will readily advertise your kennels under all circumstances. I shall have to ask the reader to overlook the apparent egotism of the statements I am now about to make, but as this book is largely the outgrowth of the author's own experience, of necessity personal matters are spoken of.

A number of years ago I received an order from the Western coast, through a Boston house, for a good all-round puppy at two hundred dollars. I sent the puppy on, and much to the surprise of the customer, stated my price for him would be one hundred instead of two. The pup matured into a very nice dog, as I expected he would, being a Cracksman pup out of a good bitch. What has been the result of this treatment? Ever since (and no later than yesterday), orders for dogs from this gentleman have been coming right along.

Another case, and this is only a sample of several from the same city: A number of years back a New York lady, accompanied by her husband, came to our kennels to purchase a dog. I had quite a handsome litter of five or six months old pups by "Merk Jr.," out of Buster stock on the dam's side, one of which, a perfectly marked seal brindle female, at once took her fancy, and she said: "We have just come from another large kennel in Boston where they asked us three hundred dollars for a little female I do not like nearly as well as this one." Her husband was one of the leading men of one of the largest trusts in the country, and money was apparently no object, and when I replied, "Mrs. Keller, that dog you select is not worth over fifty dollars (the price I afterwards sold her for) and the best dog in the litter I shall be glad to let you have for seventy-five," she seemed much surprised. I then, of course, told her that the dogs were not worth more as their muzzles were not deep enough to be worth a higher price than I wanted. I recently received a letter from her stating that her dog was still as active and much loved as ever, and the number of orders that have come to me through the sale of this dog would surprise the owners of those kennels who stick their customers with an outrageous price, and who find to their sorrow that no subsequent orders ever come, either from the customer or any one else in the vicinity. People have a way sooner or later (usually sooner) in discovering when they have been overcharged and act accordingly.

One other recommendation I wish to make in place here is: "Never try to fill an order that one has not the dogs to suit." Frankly say so, and recommend a brother fancier that you know has. One good turn deserves another and he may have a chance later to reciprocate. This creates a kindly feeling amongst kennel men, and is productive of good will, and ofttimes a large increase in business. A few years ago a lady from Connecticut came to see me to buy a first class dog or a pair, if she could get suited. I knew that in the past she had paid the highest price for her Bostons, and she wanted a dog in the neighborhood of two thousand dollars. I told her at once I had nothing for sale to suit her, but that I knew a man who owned a dog I considered worth about that sum, and recommended her strongly to buy him, and sent her to Mr. Keady, who sold to her "Gordon Boy" for that price. The sequel to this is somewhat amusing and shows how reciprocity did not take place. I went to see a litter of pups at Mr. Keady's house soon after, and expected to obtain a somewhat favorable price on the pup I picked out of the litter on account of the sale of the dog, and offered the gentleman three hundred dollars for him, upon which he replied: "Mr. Axtell, do you think that five weeks old pup is worth that sum?" and upon my replying, "I certainly do," instead of saying, "All right, take him," he exclaimed: "If that is your opinion, and I know you always say what you believe, then he is worth that sum to me," and put him back in the box. He subsequently sold him to Mr. Borden for over six thousand dollars, the highest price ever obtained for a Boston.

While writing on the subject of sales, I think it will be in order to speak of a matter that is a source of anxiety to a great many breeders, and that is the getting rid of the small bitches that are too small to breed. We have always found a ready sale for these when properly spayed for ladies' pets, largely in New York city. They make ideal house dogs, perhaps more winning and affectionate in their manner than others, never wandering off, and I believe the license fee is the same as for a male. Great care must be taken that the operation is thoroughly performed by a competent veterinary, and it is usually best done when the pup is six months old. My first experience may be of value and interest. I had a little "Buster" bitch that I felt assured to my sorrow was to small to whelp successfully, and being much fancied by a lady doctor in Waterbury, Conn., advised spaying before being sent. I took her to a veterinary with a good reputation in Boston, and after the dog had fully recovered from the operation, sent her to Dr. Conky. What was my surprise to hear that when nine months old she had come "in season." I sent the ex-President of the Boston Terrier Club, Dr. Osgood, down and an additional cost of fifty dollars ensued, whereas the first charge of two dollars would have been all that was necessary if the operation had been properly done in the first place. Am glad to say I have seen no failures since. I can conceive of no reason why there should not be a ready sale for this class of dogs in all sections of the country, and the disposal of the same will materially help the income of a great many breeders.

In conclusion let me state: "Put a price on your dogs that in your best judgment you know (not guess) to be a fair and equitable one (and if unable to decide what is right, call in an honorable expert who can) and take neither more nor less. Always remember that a man can raise horses, corn, cotton, or dogs (or any other honest product) and be a gentleman, but the moment he raises 'Cain' he ceases to be one."

Gordon Boy, Gretchen, Derby's Buster, Tommy Tucker, Ch. Lord Derby

Gordon Boy

CHAPTER XI.

BOSTON TERRIER TYPE AND THE STANDARD.

Return to Table of Contents

The standard adopted by the Boston Terrier Club in 1900 was the result of earnest, sincere, thoughtful deliberations of as conservative and conscientious a body of men as could anywhere be gotten together. Nothing was done in haste, the utmost consideration was given to every detail, and it was a thoroughly matured, and practically infallible guide to the general character and type of the breed by men who were genuine lovers of the dog for its own sake, who were perfectly familiar with the breed from its start, and who were cognizant of every point and characteristic which differentiated him from the bulldog on the one side and the bull terrier on the other, and while admitting the just claims of every other breed, believed sincerely that the dog evolved under their fostering care was the peer, if not the superior, of all in the particular sphere for which he was designed, an all-round house dog and companion. In the writer's estimation this type of dog, for the particular position in life, so to speak, he is to occupy, could not in any way be improved, and the mental qualities that accompany the physical characteristics (which are particularly specified in the first chapter) are of such inestimable value that any possible change would be detrimental. It may be observed that it was the dogs of this type that have led the van everywhere in the days when he was practically unknown outside of the state in which he originated. "Monte," "Druid Vixon," "Bonnie," "Revilo Peach," and dogs of their conformation possessed a type of interesting individuality that blazed the way east, west, north and south. Does any one imagine that the so-called terrier type one so often hears of, and which a large number of people are apparently led today to believe to be "par excellence," the correct thing, would have been capable of so doing? No one realizes more fully than the writer the fact that the bully type can be carried too far, and great harm will inevitably ensue, but the swing of the pendulum to the exaggerated terrier type will in time, I firmly believe, ring in his death knell. It is a source of wonderment to me that numbers of men who don the ermine can distribute prizes to the weedy specimens, shallow in muzzle, light in bone and substance, long in body, head and tail, who adorn (?) the shows of the past few years. I am not a prophet, neither the son of one, but I will hazard my reputation in predicting that before many years have rolled, a type, approximating that authorized by the Boston Terrier Club in 1900 will prevail, and the friends of the dog will undoubtedly believe it to be good enough to last for all time.

It will readily be recalled that Lord Byron said of the eminent actor, Sheridan, "that nature broke the die in moulding one such man," and the same may be affirmed with equal truth of the Boston terrier, and he will ever remain a type superior to and differ from all other breeds in his particular sphere.

It may not be generally known by those who are insisting on a much more terrier conformation than the standard calls for, that an equally extreme desire for an exaggerated bull type prevailed a number of years ago amongst some of the dogs' warmest supporters, whose ideal was that practically of a miniature bulldog, without the pronounced contour of the same. I remember when I joined the Club in the early days that some of the members then were afraid that the dogs were approximating too much to the terrier side of the house. What their views today would be I leave the reader to imagine. The plain fact of the case is, the dog should be a happy medium between the two, the bull and the terrier. Can any intelligent man find a chance for improvement here? I admit that many people are so constituted that a change is necessary in practically everything they are brought into close contact with. But is a change necessarily an improvement? If some men could change the color of their eyes or the general contour of their features they would never rest satisfied until they had so done, but they would speedily find out that such a change would be very detrimental to their appearance, the harmony of features and correlation of one part to another would be distorted. I admit readily that one very important result would be obtained, viz., the dog of the pronounced terrier type could be bred much more easily. But is an easy production a desideratum? I certainly think not. To those who "must be doing something" and who find a certain sense of satisfaction in tinkering with the standard, we extend our pity, and state that experience is a hard school, but some people will learn in no other. To those of us who love the dog as he is, and who believe in "letting well enough alone," we admit we might as well suggest to improve the majestic proportions of the old world cathedrals and castles we all love so much to see, or advocate the lightening up of the shadows on the canvas of the old masters, or recommend the touching up of the immortal carvings of the Italian sculptors. We advise the preacher to stick to his text, and the shoemaker to his last, and to all those who would improve the standard we say: Hands off! One very important feature in connection with the Standard is, that while breeders and judges are perfectly willing to have all dogs that come in the heavyweight class conform practically to it, when the lightweights and toys are concerned, a somewhat different type is permitted and the so-called terrier type is allowed, hence we see a tendency with the smaller dogs to a narrower chest, longer face and tail. While personally I am in favor of a dog weighing from sixteen to twenty pounds, or even somewhat heavier, there is absolutely no reason why one should not have any sized dog one desires, but please observe, do not breed small dogs at the expense of the type. Let the ten or twelve pound dog conform to the standard as much as if it weighed twenty. I think an object lesson will be of inestimable value here. Every one who has visited the poultry shows of the past few years must have been delighted and impressed to see the beautiful varieties of bantams. Take the games, for example, with their magnificent plumage and sprightly bearing. On even a casual examination it will be discovered that these little fowls are an exact reproduction of the game fowl in miniature. The same identical proportions, symmetry and shape. Take the lordly Brahma and the bantam bearing the same name, and the same exact proportions prevail. And so it should be with the small Boston terrier. They should possess the same proportions and symmetry as the larger. Remember always that when the dog is bred too much away from the bulldog type, a great loss in the loving disposition of the dog is bound to ensue. Personally, if the type had to be changed, I would rather lean to the bull type than the terrier. The following testimony of a Boston banker and director of the Union Pacific Railroad, to whom I sold two large dogs that were decidedly on the bull type, may be of interest at this point. Speaking of the first dog he said: "I have had all kinds of dogs, but I get more genuine pleasure out of my Boston terrier than all my other dogs combined. When I reach home in the afternoon I am met at the gate by Prince, and when I sit down to read my paper or a book the dog is at my feet on the rug, staying there perfectly still as long as I do. When dinner is announced he goes with me to the dining room, takes his place by my side, and every little while licks my hands, and when I go out for my usual walk before retiring the dog is waiting for me at the door while I put my hat and coat on. He follows me, never running away or barking, and he sleeps on a mat outside my door at night, and I never worry about burglars." All this is very simple and commonplace, but it shows why this type of a dog is liked. In regard to the differences of opinion that different judges exhibit when passing upon a dog in the show room, one preferring one type of a dog and the other another, this, of course, is morally wrong. The standard requirements should govern, and not individual preferences. We hear a good deal said nowadays about the cleaning up of the head, and the so-called terrier finish. That seems to be the thing to do, but does not the standard call for a compactly built dog, finished in every part of his make-up, and possessing style and a graceful carriage? This being the case, a dog should not possess wrinkled, loose skin on head or neck, and the shoulders should be neat and trim. In a word, in comporting to the standard a dog is produced that possesses a harmonious whole, "a thing of beauty" and a joy as long as he lives. In short, the dog should be as far removed from the bull type as he is from the terrier. If the present judges can not see their way clear to follow the standard, why, appoint those that will, for as every fair minded man agrees, the dogs should follow the standard and not the standard follow the dogs. It is needless to add that I do not share in the pessimistic view taken by many lovers of the dog who think he will be permanently injured by the differences of opinion that prevail as to the type, etc., and the personalities that sometimes mar the showing of the dog, for I am of the same opinion as was probably felt by the great fish who had to give up Jonah, "that it is an impossible feat to keep a good man (or dog) down," and that instead of falling off, as one writer intimates, he will fall into the good graces of a larger number of people than has heretofore fallen to the lot of any variety of man's best friend.

CHAPTER XII.

PICTURE TAKING.

Return to Table of Contents

It would seem at the first glance that to write on this subject was only a waste of time and energy, and yet I know that no one feature of the dog business is more vital in importance or more fraught with trouble than this apparently simple process of dog photography.

The novice will at once exclaim: "What could be more natural than sending on a picture of a dog I want to sell to the prospective customer? Surely he can see exactly what he is purchasing!" This may be perfectly true, and yet again it may not.

I am not writing of the subject of false pictures on the stud cards of some unscrupulous breeders, or those pictures taken of dogs whose markings are faked, only too common in some quarters. The photos look good, of course, to the buyer, but when the dog arrives, he finds, to his disgust, that the beautiful markings, in some mysterious manner, got "rubbed off" while making the journey in the crate. I recently saw a photograph of a dog sold to a Western customer, by a dealer in an adjoining town to mine, taken by an artist in photography when the dog was all "chalked up". When the dog arrived he was as free from nose band as my pocket is frequently of a dollar bill. Small wonder the buyer remarked with emphasis that the dealer was a fraud. One can almost forgive his exclamation, which he surely had not learned at Sunday school, at being taken in, in so mean a way.

I am writing more particularly of the art of the photographer in bringing out the best points of the dog, and effectually hiding the poorer ones. How many times have we heard the dealer say, in speaking of a dog with good markings, but off in many other respects: "He will make a good seller to slip away, as I can get a good looking picture of him." He knows perfectly well that a clever photographer can so pose the dog as to hide bad defects. A long muzzle, a long back, or one badly roached, poor tail, bad legs and feet, can all be minimized by posing the dog on the stand. The buyer, on receipt of the dog, although thoroughly dissatisfied, will have to admit that the photo is a genuine one, and, in most cases, is unable to obtain any redress.

Another very important side of dog photography is the mania for picture collecting. Some time ago I saw a signed article in "Dogdom", from a very charming lady living in a city fifty miles from Boston, asserting she was about to retire from the Boston terrier game, as it cost her too much to furnish photos of her dogs to people from all parts of the country, who, under the guise of wishing to buy dogs, wanted photos and pedigrees of the same. They usually stated that if they did not purchase the dog, the photo and pedigree would be promptly returned. This was the last she ever heard of them, and pictures were rarely if ever, returned. As her photos were taken by a first class photographer, the cost was considerable, and the photos were really works of art, which, perhaps, may be one reason why the recipients could not bear to let them go back. She was a lady of large wealth, and she had established a kennel of real Bostons, presided over by an expert kennel-maid, and would have become a genuine help to the breed, but "pictures" were her undoing.

Since the American dog has become the most popular breed in the canine world, many people, who cannot afford to purchase a choice specimen, seem to rest satisfied when they can obtain a photo, and they have no scruples apparently in writing to the leading kennels for pictures of their leading dogs. I have had many instances come under my notice, but, for want of space, only one typical case can be mentioned.

A few years ago, on visiting a city a short distance from Boston, I was accosted by a young man, rather flashily attired, who invited me to call and see his kennels, assuring me he had some crackerjacks. As I was unaware of the existence of any number of A-1 Bostons in his neighborhood, my curiosity was aroused and I went. I found the dogs quartered in a back room in a very small house. I have never seen such a collection of the aristocrats of the breed before or since.

When I found my voice, I managed to exclaim: "Allow me to congratulate you, my dear sir, I have never seen so many good dogs kenneled in so small a space before. You are certainly a very lucky man; the food problem never troubles you; you do not have to dodge the tax collector; no need ever to call in a vet.; no neighbors can ever complain of being kept awake at night, and the dogs that are tacked upon the ceiling seem just as content as those pasted on the walls."

He then produced his book where the pedigrees of the dogs were neatly recorded. The trouble is, he is not the only one who owns such a kennel of thorough-breds.

It must not be inferred from the above that I am averse to picture taking. By no means. They are absolutely necessary. But make them "Pen Pictures". Write a complete description of the dog in question, giving actual weight, age, conformation, color and markings, condition of health, and disposition. State the color of the brindle and the extent of the markings whether full or partial. Do not state that the dog has perfect markings if it lacks a collar or white feet. If banded only on one side of the muzzle, say so. If pinched or undershot, say so. If roached in back, poor eyes, weak in hind quarters or off in tail, say so. In fact, plainly state any defects. At the same time, if the dog is practically O. K. in all respects, stylish and trappy, do not hesitate to emphasize the fact, and if the dog likewise possesses a charming, delightful personality, make the most of it. Always remember that the perfect Boston terrier dies young!

CHAPTER XIII.

NOTES.

Return to Table of Contents

There are several features of vital import in Boston terrier breeding that the passing years have disclosed to the writer the imperative need of attention to. Most of these have been spoken of in this book before, but they seem to me at the present time to demand being specially emphasized. Feeding and its relation to skin diseases, I think, naturally heads the list.

I have received more letters of inquiry from all parts of the country asking what to do for skin trouble than for all other ailments combined. I think our little dog is more susceptible to skin affections than most dogs, owing to the fact that he is more or less a house pet, and does not get the chance of as much outdoor exercise, and the access to nature's remedy—grass, as most breeds. At the same time if fed properly, given sufficient life in the open, no dog possesses a more beautiful glossy coat.

No one factor is more responsible for skin trouble than the indiscriminate feeding of dog biscuit. These, as previously written, are first rate supplementary food, but where they are made the "piece de resistance," look out for breakers ahead. The mere fact of their being available under all circumstances and in all places contributes largely to their general use.

At the new million dollar Angell Memorial Animal Hospital, Boston, Doctors Daly and Flanigan have conducted a series of scientific experiments on dogs. I had talked with Dr. Flanigan, and stated my experience was that an exclusive dog biscuit diet was the cause of skin trouble invariably.

They selected forty dogs in perfect physical condition, dividing them into two groups of twenty each. To one was fed exclusively dog biscuits, and the other a diet of milk in the morning, and at night a feed composed of a liberal amount of spinach—they had to use the canned article as it was in winter—boiled with meat scraps and thickened with sound stale bread.

At the end of a fortnight seventeen of the first group were afflicted more or less with skin trouble, while the other twenty were in the pink of condition. To effect a cure, the spinach diet—called by the French "the broom of the stomach"—was fed, and the coat washed with a weak sulpho-naphtha solution. No internal medicine was given. In a month's time the coats of the dogs were normal. Further comment on this is unnecessary.

Next in importance to spinach I place carrots and cabbage, boiled up with the meat and rice, oat meal and occasionally corn meal. Don't be afraid to give a good quantity of the sliced boiled carrots, especially in the winter season when the dogs cannot obtain grass.

A short time ago, I went to see a group of trained monkeys and dogs perform. They both looked in beautiful condition, and on enquiring of the proprietor as to his methods of feeding, he said it was a very easy matter, as he had trained both dogs and monkeys to eat raw carrots while on the road, during which time he had to feed dog biscuits. When at home in New York he fed a vegetable hash with sound meat and rye bread, using largely carrots, beets, a very few potatoes and some apples. While on the road he had no facilities for cooking for his animals so he accustomed them to eating cut up raw carrots every other day. Previous to this he was bothered with skin trouble with both dogs and monkeys.

Champion Dean's Lady Luana

Mrs. William Kuback, with Ch. Lady Sensation

The food problem at the present time is a very serious one. The high cost of all sorts of food of every variety should force those breeders who have been keeping a very inferior stock to make up their minds once and for all that it takes just as much time and cost to raise "mutts" as it does the real article. Weed out the inferior stock that never did or will pay for their keep. Keep half a dozen good ones that will reproduce, if bred rightly, their quality, if you have not plenty of room for a large number. To those fanciers who only own two or three, sufficient food is usually furnished from the scraps left from the table, supplemented, of course, with dog biscuit.

Many kennel-men, who have a large number of dogs to feed, obtain daily from hotels or boarding houses the table scraps, and this makes an ideal food. We fed quite a large number of dogs for several years in this way with perfect success. I know of a large pack of foxhounds that are fed from the same food furnished by a large hotel. Fish heads boiled with vegetables make a good diet—be sure there are no fish hooks left in them, and the scraps from the butchers that are not quite fit for human consumption make ideal food when cooked with rice or vegetables. Be careful they are not too old, however. When skimmed milk is obtainable at the right price, with waste stale bread, it makes a well balanced ration for occasional feeding. A few onions boiled up with the feed are always in order.

I think the subject of "Tails" requires more than a passing mention here. All observers at the recent shows must have noticed the tendency toward a lengthening in many of the tails of the dogs on the bench. Some dogs have been awarded high honors which carried "more than the law allows", owing doubtless to their other excellent qualities. While I personally believe in a happy medium, never lose sight of the fact that a good short screw tail has always been, and, I believe, will always remain a leading characteristic of the American dog.

In selecting a stud dog be certain his tail is O. K. The bitch can very well afford to carry a longer one, and usually whelps better on this account. I know of nothing more discouraging in the Boston terrier game than to have a litter of choice puppies in every other respect, but off in tails.

While writing on the subject of tails, it may not be out of place to note an interesting fact in connection with this at the earliest history of our little dog. Mr. John Barnard became the possessor of Tom, afterward known as Barnard's Tom. This was the first Boston terrier to rejoice in a screw tail. Mr. Barnard did not know what to make of it, so he took the pup to old Dr. Saunders, a well known and respected veterinary surgeon of the day, to have the tail, if possible, put into splints and straightened. I guess there have been quite a number of pups, descendants of Tom, whose owners would have been only too glad to have had their straight tails put in splints, if, thereby, it would have been possible to produce a "screw".

I think the subject of sufficient importance to again call the attention of breeders to the necessity of the extreme care in breeding seal brindles. The demand started some years ago for very dark color has placed upon the market many dogs devoid of any brindle shading. At the last Boston Terrier Club specialty show a beautiful little dog, almost perfect in every other respect, was given the gate on account of being practically black.

In my former chapter on Color Breeding, I urged the necessity of using a red or light mahogany brindle on black stock. If either sex come black, never use any other color than these to mix in. Enough said!

One is constantly hearing from all parts of the country of the prevalence of bitches missing. Where they are bred to over-worked stud dogs no surprise need be manifested. In case of a "miss" have the bitch bred two or three times to the dog next time. If she misses then, the next time let her run with the dog for several days. I have written this before, but it will bear repetition.

Do not acquire the habit of getting rid of the matrons of the kennel when six or seven years old. Many bitches give birth to strong pups when eight or nine years old. I write, of course, of those in strong, vigorous condition, that have always had plenty of good outdoor exercise.

Remember, there is no spot on this broad land where the Boston terrier does not make himself thoroughly "at home." What more can one wish?

CHAPTER XIV.

CONCLUSION.

Return to Table of Contents

I was sitting by an open fire the other evening, and there passed through my mind a review of the breed since I saw a great many years ago, when the world, to me, was young, a handsome little lad leading down Beacon street, Boston, two dogs, of a different type than I had ever seen before, that seemed to have stamped upon them an individual personality and style. They were not bulldogs, neither were they bull terriers; breeds with which I had been familiar all my life; but appeared to be a happy combination of both. I need hardly say that one was Barnard's Tom, and the other his litter brother, Atkinson's Toby. Tom was the one destined to make Boston terrier history, as he was the sire of Barnard's Mike.

Mr. J. P. Barnard has rightly been called the "Father of the Boston terrier," and he still lives, hale and hearty. May his last days be his best, and full of good cheer!

I am now rapidly approaching the allotted time for man, but I venture the assertion that were I to visit any city or even small town of the United States or Canada, I could see some handsome little lad or lassie leading one of Barnard's Mike's sons or daughters. Small wonder he is called the American dog.

The celebrated Dr. Johnson once remarked that few children live to fulfil the promise of their youth. Our little aristocrat of the dog world has more than done so. May his shadow never grow less!

I feel convinced that I ought to take this opportunity to record my kindly appreciation of the generous expressions of thanks for my efforts on behalf of the dog. They have come from all parts of the country, and from all classes of people. Were it in my power I would gladly reply to each individual writer. This is impossible. I can only say, "I thank you! May God bless us, one and all!"

CHAPTER XV.

TECHNICAL TERMS USED IN RELATION TO THE BOSTON TERRIER, AND THEIR MEANING.

Return to Table of Contents

- A Crackerjack — A first class, typical dog.
- A Mutt — A worthless specimen.
- A Flyer — A dog capable of winning in any company.
- A Weed — A leggy, thin, attenuated dog, bred so.
- A Fake — A dog whose natural appearance has been interfered with to hide defects.
- A Dope — A dog afflicted, usually with chorea, that has had cocaine administered to him to stop the twitching while in the judging ring.
- A Ringer — A dog shown under a false name, that has previously been shown under his right name.
- Apple-headed — Skull round, instead of flat on top.
- Broken-up Face — Bulldog face, with deep stop and wrinkle and receding nose.
- Frog or Down Face — Nose not receding.
- Dish-faced — One whose nasal bone is higher at the nose than at the stop.
- Butterfly Nose — A spotted nose.
- Dudley Nose — A flesh-colored nose.
- Rose Ear — An ear which the tip turns backward and downward, disclosing the inside.
- Button Ear — An ear that falls over in front, concealing the inside.
- Tulip Ear — An upright, or pricked ear.
- Blaze — The white line up the face.
- Cheeky — When the cheek bumps are strongly defined.
- Occiput — The prominent bone at the back or top of the skull, noticeably prominent in bloodhounds.
- Chops — The pendulous lips of the bulldog.
- Cushion — Fullness in the top lips.

- Dewlap — The pendulous skin under the throat.
- Lippy — The hanging lips of some dogs, who should not possess same, as in the bull terrier.
- Layback — A receding nose.
- Pig-jawed — The upper jaw protruding over the lower; an exaggeration of an undershot jaw.
- Overshot — The upper teeth projecting beyond the lower.
- Undershot — The lower incisor teeth projecting beyond the upper, as in bulldogs.
- Wrinkle — Loose, folding skin over the skull.
- Wall Eye — A blue mottled eye.
- Snipy — Too pointed in muzzle; pinched.
- Stop — The indentation between the skull and the nasal bone near the eyes.
- Septum — The division between the nostrils.
- Leather — The skin of the ear.
- Expression — The size and placement of the eye determines the expression of the dog.
- Brisket — That part of the body in front of the chest and below the neck.
- Chest — That part of the body between the forelegs, sometimes called the breast, extending from the brisket to the body.
- Cobby — Thick set; low in stature, and short coupled; or well ribbed up, short and compact.
- Couplings — The space between the tops of the shoulder blades, and the tops of the hip joints. A dog is accordingly said to be long or short "in the couplings."
- Deep in Brisket — Deep in chest.
- Elbows — The joint at the top of forearm.
- Elbows Out — Self-explanatory; either congenital, or as a result of weakness.
- Flat-sided — Flat in ribs; not rounded.
- Forearm — The foreleg between the elbows and pastern.
- Pastern — The lower section of the leg below the knee or hock respectively.

- Shoulders — The top of the shoulder blades, the point at which a dog is measured.
- Racy — Slight in build and leggy.
- Roach-back — The arched or wheel formation of loin.
- Pad — The underneath portion of the foot.
- Loins — The part of body between the last rib and hindquarters.
- Long in flank — Long in back of loins.
- Lumber — Unnecessary flesh.
- Cat-foot — A short, round foot, with the knuckles well developed.
- Hare-foot — A long, narrow foot, carried forward.
- Splay-foot — A flat, awkward forefoot, usually turned outward.
- Stifles — The upper joint of hind legs.
- Second Thighs — The muscular development between stifle joint and hock.
- The Hock — The lowest point of the hind leg.
- Spring — Round, or well sprung ribs; not flat.
- Shelly — Narrow, shelly body.
- Timber — Bone.
- Tucked Up — Tucked up loin, as seen in greyhounds.
- Upright Shoulders — Shoulders that are set in an upright, instead of an oblique position.
- Leggy — Having the legs too long in proportion to body.
- Stern — Tail.
- Screw Tail — A tail twisted in the form of a screw.
- Kink Tail — A tail with a break or kink in it.
- Even Mouthed — A term used to describe a dog whose jaws are neither overhung nor underhung.
- Beefy — Big, beefy hind quarters.
- Bully — Where the dog approaches the bulldog too much in conformation.
- Terrier Type — Where the dog approaches the terrier too much in conformation.
- Cow-hocked — The hocks turning inward.

- Saddle-back—The opposite of roach-back.
- Lengthy—Possessing length of body.
- Broody—A broody bitch; one whose length of conformation evidences a likely mother; one who will whelp easily and rear her pups.
- Blood—A blood; a dog whose appearance denotes high breeding.
- Condition—Another name for perfect health, without superfluous flesh, coat in the best of shape, and spirits lively and cheerful.
- Style—Showy, and of a stylish, gay demeanor.
- Listless—Dull and sluggish.
- Character—A sub-total of all the points which give to the dog the desired character associated with his particular variety, which differentiates him from all other breeds.
- Hall-mark—That stamp of quality that distinguishes him from inferior dogs, as the sterling mark on silver, or the hall-mark on the same metal in England.

www.ingramcontent.com/pod-product-compliance
Lightning Source LLC
Chambersburg PA
CBHW031419210526
45464CB00005B/1956